Supporting open minds since 2005

Recent Advances in Polynomials
http://dx.doi.org/10.5772/intechopen.95161
Edited by Kamal Shah

Contributors
Lhoussain El Fadil, Mohamed Faris, Mritunjay Kumar Singh, Rajesh P. Singh, Jerome A. Adepoju, Sándor Kovács, Szilvia György, Noémi Gyúró, Mehmet Firat, Emre Esener, Toros Arda Akşen, Bora Şener, Shivani Verma, Mani Shankar Prasad, Kamal Shah, Okba Weslati, Samir Bouaziz, Mohamed-Moncef Serbaji, Eiman, Hammad Khalil, Rahmat Ali Khan, Thabet Abdeljawad

First published in London, United Kingdom, 2022 by IntechOpen
IntechOpen is the global imprint of INTECHOPEN LIMITED, registered in England and Wales, registration number: 11086078, 5 Princes Gate Court, London, SW7 2QJ, United Kingdom
Printed in Croatia

British Library Cataloguing-in-Publication Data
A catalogue record for this book is available from the British Library

Additional hard and PDF copies can be obtained from orders@intechopen.com

Recent Advances in Polynomials
Edited by Kamal Shah
p. cm.
Print ISBN 978-1-83969-758-6
Online ISBN 978-1-83969-759-3
eBook (PDF) ISBN 978-1-83969-760-9

For EU product safety concerns:
IN TECH d.o.o., Prolaz Marije Krucifikse Kozulić 3, 51000 Rijeka, Croatia, info@intechopen.com or visit our website at intechopen.com.

Recent Advances in Polynomials

Edited by Kamal Shah

Published in London, United Kingdom

IntechOpen

Meet the editor

Dr. Kamal Shah is a senior researcher in the Department of Mathematics & Sciences, Prince Sultan University, Riyadh, Saudi Arabia. From 2007 to 2021, he was an associate professor in the Department of Mathematics, University of Malakand, Chakdara, Pakistan. He obtained a master's in Mathematics from Government Post Graduate Jahanzeb College, Swat, Pakistan, in 2005. He joined the University of Malakand as a lecturer in 2007 and obtained his Ph.D. in Fractional Calculus in 2016 from the same university. Dr. Shah has published numerous articles in scientific journals and has supervised many MPhil and Ph.D. students.

Contents

Preface

A polynomial in mathematics is an expression consisting of coefficients and variables that involves only the operations of addition, subtraction, and multiplication with a non-negative integer power. Polynomials have significant applications in the description of many real-world problems. Various polynomial-type expressions are being used to describe the various chemical, biological, social, and economic problems. Further, polynomials are increasingly being used in numerical computations of large numbers of nonlinear problems. They are also used in calculus and numerical analysis to approximate other functions. In linear algebra during spectral analysis, eigen values are computed through characteristic polynomials, which further help in finding the radius of convergence of various matrices. In stability analysis, eigen values are computed through the help of minimal polynomials of the Jacobian matrix. Polynomials are powerful tools in approximation theory and advanced numerical analysis.

Chapter 1 discusses some characteristic functions and various valuable relations of polynomials.

Chapter 2 presents some problems of permutations and their applications, various relations, and results.

Chapter 3 discusses the effectiveness of basic sets of Gončarov polynomials and their different properties.

Chapter 4 reviews irreducible factors of polynomials as well as discusses the irreducibility of polynomials with specific requirements on their coefficients.

Chapter 5 describes the use of homogenous polynomials yield function and various results.

Chapter 6 examines the irreducibility of polynomials in non-binary fields.

Chapter 7 presents the efficiency of polynomial regression algorithms.

Chapter 8 describes the use of shifted Jacobi polynomials in some fractional order differential equations under initial and boundary conditions.

Kamal Shah
Department of Mathematics and Sciences,
Prince Sultan University,
Riyadh, Saudi Arabia

Department of Mathematics,
University of Malakand,
Chakdara, Pakistan

Chapter 1

Characteristic Polynomials

Sándor Kovács, Szilvia György and Noémi Gyúró

Abstract

In this chapter, we provide a short overview of the stability properties of polynomials and quasi-polynomials. They appear typically in stability investigations of equilibria of ordinary and retarded differential equations. In the case of ordinary differential equations we discuss the Hurwitz criterion, and its simplified version, the Lineard-Chippart criterion, furthermore the Mikhailov criterion and we show how one can prove the change of stability via the knowledge of the coefficients of the characteristic polynomial of the Jacobian of the given autonomous system. In the case of the retarded differential equation we use the Mikhailov criterion in order to estimate the length of the delay for which no stability switching occurs. These results are applied to the stability and Hopf bifurcation of an equilibrium solution of a system of ordinary differential equations as well as of retarded dynamical systems.

Keywords: Hurwitz stability, Schur stability, Mikhailov criterion, Hopf bifurcation

1. Introduction

As it is well-known, many systems of applied mathematics are modeled by retarded functional differential equations of type

$$\dot{x} = f(x, x(\cdot - \tau)) \tag{1}$$

(cf. [1]), where $f : \Omega \times \Omega \to \mathbb{R}^d$ is continuously differentiable and $\Omega \subset \mathbb{R}^d$ is an open set. Here $\tau \geq 0$ represents the so-called delay or time lag. In order to have a solution in some interval $(0, r)$, $r > 0$ one has to know the solution on $[-\tau, 0]$, which means that one has to attach a continuous initial function $\phi : [-\tau, 0] \to \mathbb{R}^d$ as an initial condition to the system (cf. [2]). Clearly, in the case of $\tau = 0$ one has to deal with the initial value problem for ordinary differential equations.

In order to examine the stability of an equilibrium $a \in \Omega$ of system (1), i.e. the equilibrium solution

$$\hat{a} : \mathbb{R} \to \mathbb{R}, \qquad \hat{a}(t) := a \tag{2}$$

for which $f(a, a) = 0$ holds, one has to discuss the spectral properties of the linearized system

$$\dot{u} = Au + Bu(\cdot - \tau) \tag{3}$$

where

$$A := \partial_1 f(a, a) \in \mathbb{R}^{d \times d} \quad \text{and} \quad B := \partial_2 f(a, a) \in \mathbb{R}^{d \times d}. \tag{4}$$

It may be supposed that system (3) has a solution of the form $\mathbb{R} \ni t \mapsto e^{\lambda t} \cdot s$ where $0 \neq s \in \mathbb{R}^d$, that is

$$\left(\lambda I_d - A - Be^{-\lambda\tau}\right)s = 0. \tag{5}$$

This can only happen if and only if $\Delta(\lambda; \tau) = 0$ holds, where

$$\Delta(z; \tau) := \det(zI_d - A - Be^{-z\tau}) \qquad (z \in \mathbb{C}) \tag{6}$$

is called characteristic quasi-polynomial of the linear delay system (3).

The organization of the chapter is as follows. In the next section, we introduce and prove two criteria regarding stability of $\Delta := \Delta(\cdot; 0)$, i.e. we give conditions for which the zeros of Δ have negative real parts. Concretely, we deal with the Hurwitz criterion and with its simplified version, the Lineard-Chippart criterion, and the Mikhailov criterion. Furthermore, we show how one can check the conditions of Hopf bifurcation via knowledge of the coefficients of the characteristic polynomial. In the section that follows we examine the case when the delay τ is positive. We show how the Mikhailov criterion can be extended for quasi-polynomials and how the length of the delay can be estimated in order to have stability. Finally, we present a criterion for Hopf bifurcation.

2. The undelayed case: $\tau = 0$

If there is no delay present, i.e. $\tau = 0$ holds then we have to deal with the characteristic polynomial

$$\chi(z) := (-1)^d \det(A - zI_d) = z^d + a_{d-1}z^{d-1} + a_{d-2}z^{d-2} + \dots + a_1 z + a_0 \qquad (z \in \mathbb{C}) \tag{7}$$

where the coefficients of χ_A are determined recursively by the Faddeev–LeVerrier algorithm (cf. [3, 4]) as follows

$$\left.\begin{aligned} &N_0 := O && a_d := 1 && (k = 0), \\ &N_k := AN_{k-1} + a_{d-k+1}I_d && a_{d-k} := -\frac{1}{k}\mathrm{Tr}(AN_k) && (k \in \{1, \dots, d\}). \end{aligned}\right\} \tag{8}$$

2.1 The stability of the characteristic polynomial χ_A

The asymptotic stability of (3) is determined by the stability of the matrix A, i.e. by the stability of its characteristic polynomial χ_A. We are now supplying some criteria for the stability of the characteristic polynomial χ_A. Under stability, we mean the so-called Hurwitz stability, i.e. the zeros of χ_A lie in the open left half of the complex plane. In this case χ_A is called also Hurwitz polynomial.

There is a very simple but very important necessary condition for χ_A to be Hurwitz (cf. [5, 6]).

Theorem 1.1 (Stodola). If the characteristic polynomial χ_A in (7) is stable then all of its coefficients are positive, i.e. $a_k > 0$ holds where $k \in \{0 \dots, d-1\}$.

Proof: The real and complex zeros of χ_A my be written as λ_k, resp. $\alpha_l \pm i\beta_l$, where α_l and β_l are both real. If the multiplicity of the real, resp. complex zeros are denoted by σ_k, resp. τ_l, where $k \in \{1, \dots, r\}$, resp. $l \in \{1, \dots, s\}$, then

$$\sum_{k=1}^{r} \sigma_k + 2\sum_{l=1}^{s} \tau_l = d, \tag{9}$$

and we can split χ_A into linear, resp. quadratic factors according to the real, resp. complex zeros as follows

$$\begin{aligned} \chi_A(z) &= a_0 + a_1 z + \dots + a_{d-1} z^{d-1} + z^d \\ &= \prod_{k=1}^{r} (z - \lambda_k)^{\sigma_k} \cdot \prod_{l=1}^{s} (z - \alpha_l - i\beta_l)^{\tau_l} (z - \alpha_l + i\beta_l)^{\tau_l} \\ &= \prod_{k=1}^{r} (z - \lambda_k)^{\sigma_k} \cdot \prod_{l=1}^{s} \left(z^2 - 2\alpha_l z + \alpha_l^2 + \beta_l^2\right)^{\tau_l}. \end{aligned} \tag{10}$$

Thus, the stability of χ_A implies the sign conditions

$$\lambda_k < 0, \quad \alpha_l < 0 \qquad (k \in \{1, \dots, r\};\ l \in \{1, \dots, s\}). \tag{11}$$

This means that all coefficients of all factors in the product above are positive. By performing the multiplications one can see that the coefficients of χ_A are positive. ■

In the case of $d = 1$ and $d = 2$ this criterion is sufficient and necessary. Indeed, in case of $d = 1$ the characteristic polynomial has the form

$$\chi_A(z) = z + a_0 \qquad (z \in \mathbb{C}), \tag{12}$$

furthermore in case of $d = 2$ we have

$$a_1 = -\mathrm{Tr}(A), \qquad a_0 = -\frac{1}{2}\left\{\mathrm{Tr}(A^2) - \mathrm{Tr}(A)^2\right\} = \det(A) \tag{13}$$

(cf. (8)), thus

$$\chi_A(\xi_\pm) = 0 \iff \xi_\pm = \frac{\mathrm{Tr}(A) \pm \sqrt{\mathrm{Tr}(A)^2 - 4\det(A)}}{2} \tag{14}$$

and χ_A is stable if and only if $\mathrm{Tr}(A) < 0$ and $\det(A) > 0$ hold, because if

- $\mathrm{Tr}(A)^2 - 4\det(A) > 0$, then we have $\xi_- < 0$,

$$\xi_+ < 0 \iff \mathrm{Tr}(A)^2 - 4\det(A) < \mathrm{Tr}(A)^2 \iff \det(A) > 0; \tag{15}$$

- $\mathrm{Tr}(A)^2 - 4\det(A) = 0$, then the zeros ξ_- and ξ_+ are equal and real:

$$\xi_- = \xi_+ = -\frac{\mathrm{Tr}(A)}{2} < 0 \iff \mathrm{Tr}(A) > 0; \tag{16}$$

- $\mathrm{Tr}(A)^2 - 4\det(A) < 0$, then there are only complex zeros:

$$\Re(\xi_\pm) = -\frac{\mathrm{Tr}(A)}{2} < 0 \iff \mathrm{Tr}(A) > 0. \tag{17}$$

Unfortunately the criterion is for $d > 2$ not sufficient. For example, the polynomial

$$p(z) := z^4 + 3z^3 + 3z^2 + 3z + 3 = (z^2 + 1)(z + 1)(z + 2) \qquad (z \in \mathbb{C}) \tag{18}$$

has positive coefficients, but two of its zero, namely $\pm i$ are not in the open left half-plane. In case of $d = 3$ there is a result which can be proved in several ways. Pontryagin (cf. [7]) proves it in a circumstantial way. He uses that the zeros of a polynomial are continuous functions of the coefficients (cf. [8, 9]). Our presentation is based on the results of Suter (cf. [10]).

Theorem 1.2 In case of $d = 3$ the characteristic polynomial of A has the form

$$\chi_A(z) := z^3 - \mathrm{Tr}(A)z^2 + \mathrm{Tr}(\mathrm{adj}(A))z - \det(A) \qquad (z \in \mathbb{C}) \tag{19}$$

(cf. (8)) and χ_A is stable if and only if

$$\mathrm{Tr}(A) < 0, \qquad \det(A) < 0, \qquad \mathrm{Tr}(A) \cdot \mathrm{Tr}(\mathrm{adj}(A)) < -\det(A) \tag{20}$$

hold

Proof: Using the Faddeev-LeVerrier-algorithm (cf. (8)) we have

$$\chi_A(z) = z^3 + az^2 + bz + c \qquad (z \in \mathbb{C}), \tag{21}$$

where

$$a := -\mathrm{Tr}(A), \qquad b := \frac{\mathrm{Tr}(A)^2 - \mathrm{Tr}(A^2)}{2} = \mathrm{Tr}(\mathrm{adj}(A)), \qquad c := -\det(A). \tag{22}$$

As a consequence of the fundamental theorem of algebra, we can split χ_A into a linear and a quadratic factor

$$\chi_A(z) = (z - \alpha)(z^2 + \beta z + \gamma) = z^3 + (\beta - \alpha)z^2 + (\gamma - \alpha\beta)z - \alpha\gamma \qquad (z \in \mathbb{C}). \tag{23}$$

In view of the above considerations for the first and the second-order polynomials we see that χ_A is stable if and only if $\alpha < 0$, $\beta > 0$ and $\gamma > 0$ hold. Thus, it is enough to show that the equivalence

$$\alpha < 0, \ \beta > 0, \ \gamma > 0 \iff a > 0, \ c > 0, \ ab - c > 0. \tag{24}$$

holds. We prove this statement in two steps.

Step 1. We prove that the positivity of the coefficients a, b, c entails

$$\alpha < 0, \quad \gamma > 0 \quad \text{and} \quad \mathrm{sgn}(\beta) = \mathrm{sgn}(ab - c). \tag{25}$$

Indeed,

- from $-\alpha\gamma = c > 0$ it follows that $\alpha \neq 0$, $\gamma \neq 0$. Hence the equivalence

$$\alpha > 0 \iff \gamma < 0 \tag{26}$$

holds. The case $\alpha > 0, \gamma < 0$ cannot happen, because $\gamma - \alpha\beta = b > 0$ would imply $\beta < 0$, which is not possible due to $\beta - \alpha = a > 0$.

$$\alpha^2 + b > 0 \tag{27}$$

$$ab - c = (\beta - \alpha)(\gamma - \alpha\beta) + \alpha\gamma = \alpha^2\beta - \alpha\beta^2 + \beta\gamma = \beta(\alpha^2 - \alpha\beta + \gamma) = \\ = \beta(\alpha^2 + b). \quad (28)$$

Step 2. It remains to prove that the equivalence (24) holds.

- If χ_A is stable, i.e. $\alpha < 0, \beta > 0$ and $\gamma > 0$ hold, then $a > 0, b > 0$ and clearly $c > 0$, even by Step 2 $\operatorname{sgn}(ab - c) = \operatorname{sgn}(\beta) = 1$, i.e. $ab - c > 0$.

- If inequalities $a > 0, c > 0$ and $ab - c > 0$ hold, then $b > 0$, hence by Step 2 we have $\alpha < 0, \gamma > 0$ and $\operatorname{sgn}(\beta) = \operatorname{sgn}(ab - c) = 1$, i.e. $\beta > 0$ which completes the proof. ■

Example 1. If the matrix $A \in \mathbb{R}^{3\times 3}$ is antisymmetric, i.e. for suitable $a, b, c \in \mathbb{R}$

$$A = \begin{bmatrix} 0 & a & b \\ -a & 0 & c \\ -b & -c & 0 \end{bmatrix} \quad (29)$$

holds, then its characteristic polynomial has the form

$$\chi_A(z) = z^3 + (a^2 + b^2 + c^2)z = z\left[z^2 + (a^2 + b^2 + c^2)\right] \qquad (z \in \mathbb{C}). \quad (30)$$

This means that χ_A and hence A is unstable.

In order to formulate the necessary and sufficient stability condition for the polynomial χ_A with arbitrary degree $d \in \mathbb{N}$, we shall first fix our terminology. Let us define the Hurwitz matrix of the characteristic polynomial χ_A by

$$\mathcal{H}_{\chi_A} := [h_{ij}], \quad \text{where} \quad h_{ij} := \begin{cases} a_{d-(2j-i)} & (0 \le 2j - i \le d), \\ 0 & (\text{elsewhere}) \end{cases} \qquad (i,j \in \{1, \dots, d\}), \quad (31)$$

i.e.

$$\mathcal{H}_{\chi_A} := \begin{bmatrix} a_{d-1} & a_{d-3} & a_{d-5} & \dots & a_{d-2d+3} & a_{d-2d+1} \\ a_d & a_{d-2} & a_{d-4} & \dots & a_{d-2d+4} & a_{d-2d+2} \\ 0 & a_{d-1} & a_{d-3} & \dots & a_{d-2d+5} & a_{d-2d+3} \\ 0 & a_d & a_{d-2} & \dots & a_{d-2d+6} & a_{d-2d+4} \\ \vdots & & & \ddots & & \vdots \\ 0 & 0 & \dots & \dots & a_1 & 0 \\ 0 & 0 & \dots & \dots & a_2 & a_0 \end{bmatrix} \in \mathbb{R}^{d\times d} \quad (32)$$

where $a_d := 1$ and $a_{-m} := 0$ if $m > 0$.

Theorem 1.3 (Routh-Hurwitz criterion). The characteristic polynomial χ_A in (7) is stable if and only if all leading principal minors

$$\Delta_k := \det \begin{bmatrix} h_{11} & \dots & h_{1k} \\ \vdots & \ddots & \vdots \\ h_{k1} & \dots & h_{kk} \end{bmatrix} \qquad (k \in \{1, \dots, d\}) \quad (33)$$

of $\mathcal{H}_{\chi_A}$ are positive, i.e.

$$\Delta_1 = a_{d-1} > 0, \quad \Delta_2 = \det\begin{bmatrix} a_{d-1} & a_{d-3} \\ a_d & a_{d-2} \end{bmatrix} > 0, \quad \ldots, \quad \Delta_d = a_0 \Delta_{d-1} > 0 \tag{34}$$

hold. □

For

- two-dimensional system we have

$$\mathcal{H}_{\chi_A} := \begin{bmatrix} a_1 & a_{-1} \\ 1 & a_0 \end{bmatrix} = \begin{bmatrix} a_1 & 0 \\ 1 & a_0 \end{bmatrix}. \tag{35}$$

Thus, this criterion can be stated as

$$\Delta_1 = a_1 > 0, \qquad \Delta_2 = a_1 - a_0 > 0, \tag{36}$$

or

$$a_0 > 0, \qquad a_1 > 0. \tag{37}$$

- third-dimensional system we have

$$\mathcal{H}_{\chi_A} := \begin{bmatrix} a_2 & a_0 & 0 \\ 1 & a_1 & 0 \\ 0 & a_2 & a_0 \end{bmatrix}. \tag{38}$$

Thus, this criterion is

$$\Delta_1 = a_2 > 0, \qquad \Delta_2 = a_2 a_1 - a_0 > 0, \qquad \Delta_3 = a_0 \Delta_2 > 0, \tag{39}$$

or

$$a_0 > 0, \qquad a_2 > 0, \qquad a_2 a_1 > a_0. \tag{40}$$

- fourth-dimensional system we have

$$\mathcal{H}_{\chi_A} := \begin{bmatrix} a_3 & a_1 & 0 & 0 \\ 1 & a_2 & a_0 & 0 \\ 0 & a_3 & a_1 & 0 \\ 0 & 1 & a_2 & a_0 \end{bmatrix}. \tag{41}$$

Thus, this criterion can be stated as $\Delta_1 = a_3 > 0$,

$$\Delta_2 = a_3 a_2 - a_1 > 0, \quad \Delta_3 = a_3 a_2 a_1 - a_3^2 a_0 - a_1^2 > 0, \quad \Delta_4 = a_0 \Delta_3 > 0 \tag{42}$$

or

$$a_0 > 0, \qquad a_3 > 0, \qquad a_3 a_2 > a_1, \qquad a_1 a_2 a_3 > a_0 a_3^2 + a_1^2. \tag{43}$$

As the application of the above theorem, we mention the Orlando formula (cf. [11]) which establishes the useful relation between the Hurwitz determinants and the polynomial whose roots are sums of the roots of a given polynomial and which can be proved by mathematical induction (cf. [12]).

Theorem 1.4 (Orlando-formula) If $\xi_1, \dots, \xi_d$ are the roots of the characteristic polynomial χ_A then the $(d-1)$-th principal minor of the Hurwitz matrix can be expressed as

$$\Delta_{d-1} = (-1)^{d(d-1)/2} \cdot \prod_{\substack{i,j=1 \\ i<j}}^{d} \left(\lambda_i + \lambda_j\right). \tag{44}$$

In case of

- $d = 2$ this formula reduces to the well known Vieta formula in the quadratic equation

$$a_1 = \Delta_1 = -(\lambda_1 + \lambda_2); \tag{45}$$

- $d = 3$ the formula in (44) reduces to

$$a_2 a_1 - a_0 = \Delta_2 = -(\lambda_1 + \lambda_2)(\lambda_1 + \lambda_3)(\lambda_2 + \lambda_3). \tag{46}$$

We remark (cf. [12]) if criterion (34) is satisfied then χ_A is stable which due to the form (10) has the consequence that all coefficients of χ_A are positive:

$$a_0 > 0, \qquad a_1 > 0, \qquad \dots, \qquad a_{d-1} > 0. \tag{47}$$

Clearly, if (47) holds then condition (34) is redundant: many of inequalities in (34) are unnecessary. For χ_A to be a Hurwitz stable, a necessary and sufficient condition can be established which requires about half amount of computations needed in the criterion of Routh-Hurwitz (cf. [12, 13]).

Theorem 1.5 (Liénard-Chipart) The following statements are equivalent:

1. the characteristic polynomial χ_A in (7) is Hurwitz stable;
2. $a_0 > 0,\ a_2 > 0,\ \dots;\ \Delta_1 > 0,\ \Delta_3 > 0,\ \dots;$
3. $a_0 > 0,\ a_2 > 0,\ \dots;\ \Delta_2 > 0,\ \Delta_4 > 0,\ \dots;$
4. $a_0 > 0,\ a_1 > 0,\ a_3 > 0\ \dots;\ \Delta_1 > 0,\ \Delta_3 > 0,\ \dots;$
5. $a_0 > 0,\ a_1 > 0,\ a_3 > 0\ \dots;\ \Delta_2 > 0,\ \Delta_4 > 0,\ \dots.$

Example 2. For $\alpha, \beta \in \mathbb{R}$ the matrix

$$A := \begin{bmatrix} 0 & 1 & 0 & 0 & 0 \\ 0 & 0 & 1 & 0 & 0 \\ 0 & 0 & 0 & 1 & 0 \\ 0 & 0 & 0 & 0 & 1 \\ \alpha\beta - 1 & -2 & -3 & -2 & -1 \end{bmatrix} \tag{48}$$

has the characteristic polynomial

$$\chi_A(z) := 1 - \alpha\beta + 2z + 3z^2 + 2z^3 + z^4 \qquad (z \in \mathbb{C}). \tag{49}$$

It suffices to calculate

$$\Delta_4 = \det\left(\mathcal{H}_{\chi_A}\right) = (1-\alpha\beta)\Delta_3 = (1-\alpha\beta)\cdot\det\begin{bmatrix} 2 & 2 & 0 \\ 1 & 3 & 1-\alpha\beta \\ 0 & 2 & 2 \end{bmatrix} \tag{50}$$

$$= (1-\alpha\beta)\cdot\{12 - 4(1-\alpha\beta) - 4\} = 4(1-\alpha\beta)(1+\alpha\beta).$$

Thus, χ_A and hence A is stable if and only if

$$(1-\alpha\beta > 0 \;\text{ and }\; 1+\alpha\beta > 0), \quad \text{i.e.} \quad -1 < \alpha\beta < 1 \tag{51}$$

holds (cf. **Figure 1**).

There is a criterion of geometric character which is useful for the study of the stability of χ_A.

Definition 1. *The curve*

$$\Gamma_A := \{\chi_A(i\omega) = (\Re(\chi_A(i\omega)), \Im(\chi_A(i\omega))) \in \mathbb{C} : \; \omega \in \mathbb{R}\} \tag{52}$$

is called Mikhailov hodograph or amplitude-phase curve (cf. [14]).

Some geometrical properties of the hodograph Γ_A are in strong relationship to the stability of the characteristic polynomial χ_A. This polynomial has no zero on the imaginary axis if and only if the origin does not lie on the curve Γ_A. In this case the function

$$\varphi_{\chi_A}(\omega) := \arg(\chi_A(i\omega)) = \Im(\log(\chi_A(i\omega))) \qquad (\omega \in \mathbb{R}) \tag{53}$$

is continuous in every point of the real line. Moreover, we deal with the change

$$\Delta_{\chi_A} := \Delta_{\omega\in(-\infty,\infty)}\varphi_{\chi_A}(\omega) \tag{54}$$

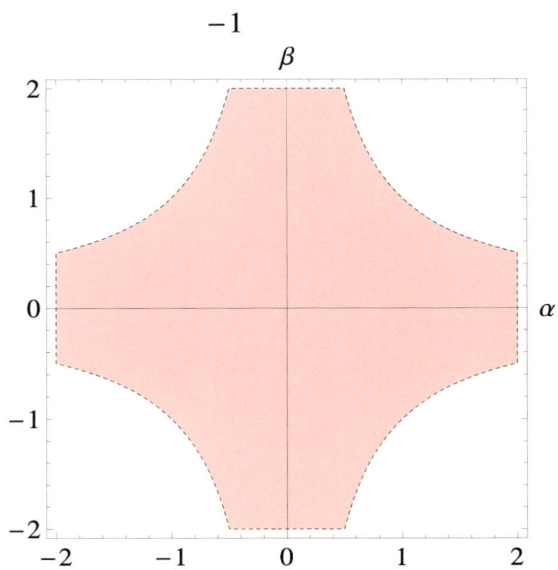

Figure 1.
Stability chart of the polynomial $\mathbb{C} \ni z \mapsto z^4 + 2z^3 + 3z^2 + 2z + 1 - \alpha\beta$.

where Δ_{χ_A} denotes the change of argument of the vector $\varphi_{\chi_A}(\omega)$ in the complex plane as ω increases from $-\infty$ to $+\infty$. Because

$$\chi_A(-i\omega) = \overline{\chi_A(i\omega)}, \tag{55}$$

therefore we have

$$\Delta_{\omega\in[0,\infty)}\varphi_{\chi_A}(\omega) = \frac{1}{2}\cdot\Delta_{\omega\in(-\infty,\infty)}\varphi_{\chi_A}(\omega). \tag{56}$$

This means that it is enough to know the behavior of the vector $\varphi_{\chi_A}(\omega)$ for $0 \leq \omega < \infty$. The next theorem which is known as the Mikhailov criterion of stability is based on the principle of argument. Because the form as it is in the next theorem is a special case of the one formulated in the next section we omit its proof now.

Theorem 1.6 (Mikhailov). The polynomial χ_A is stable if and only if the following two conditions are fulfilled:

1. the curve Γ_A does not cross the origin, i.e. the implication

   $$\chi_A(z) = 0 \;\Rightarrow\; \Re(z) \neq 0$$

   is true, which means that χ_A has no zeros on the imaginary axis;

2. the curve Γ_A encircles the origin anticlockwise at an angle $d\pi/2$ while ω changes from 0 to $+\infty$, i.e.

   $$\Delta_{\omega\in[0,\infty)}\arg(\chi_A(i\omega)) = \frac{d\pi}{2}$$

   holds.

Example 3. In case of $d = 2$ we have for the characteristic polynomial

$$\chi_A(z) := a_0 + a_1 z + z^2 \qquad (z\in\mathbb{C}): \tag{57}$$

$$\Re(\chi_A(i\omega)) = a_0 - \omega^2, \quad \Im(\chi_A(i\omega)) = a_1\omega, \tag{58}$$

from which

$$\sin(\arg(\chi_A(i\omega))) = \frac{\Im(\chi_A(i\omega))}{\sqrt{[\Re(\chi_A(i\omega))]^2 + [\Im(p(i\omega))]^2}} = \frac{a_1\omega}{\sqrt{[a_0-\omega^2]^2 + a_1^2\omega^2}} \tag{59}$$

$$\to \operatorname{sgn}(a_1)0 \quad (\omega \to +\infty),$$

resp. as $\omega \to +\infty$

$$\cos(\arg(\chi_A(i\omega))) = \frac{\Re(\chi_A(i\omega))}{\sqrt{[\Re(\chi_A(i\omega))]^2 + [\Im(\chi_A(i\omega))]^2}} = \frac{a_0-\omega^2}{\sqrt{[a_0-\omega^2]^2 + a_1^2\omega^2}} \to -1 \tag{60}$$

follows. Thus,

$$\Delta_{\omega\in[0,+\infty)}\arg(\chi_A(i\omega)) = \begin{cases} 0 & (a_1<0 \wedge a_0<0), \\ -\pi & (a_1<0 \wedge a_0>0), \\ 0 & (a_1>0 \wedge a_0<0), \\ \pi & (a_1>0 \wedge a_0>0) \end{cases} \tag{61}$$

which means that χ_A is stable if and only if $a_0 > 0$ and $a_1 > 0$ hold.

Often what is to be checked is not the stability of the characteristic polynomial χ_A but the question as to whether every zero of the polynomial lies in the interior of the unit circle around the origin of the complex plane. In this case χ_A is called Schur stable polynomial or simply Schur polynomial. This phenomenon plays a crucial role in the stability of discrete dynamical systems and in the asymptotic stability of periodic linear systems (cf. [15]). Regarding this problem there are two main treatments. The first way to investigate the Schur stability of χ_A is to introduce the Möbius-transformation

$$w := \frac{z+1}{z-1} \quad (-1 \neq z \in \mathbb{C}) \qquad /z := \frac{w+1}{w-1} \quad (1 \neq w \in \mathbb{C})/ \tag{62}$$

which takes the interior of the unit circle of the complex plane $\{z \in \mathbb{C} : |z| < 1\}$ into the interior of the left half-plane $\{w \in \mathbb{C} : \mathfrak{R}(w) < 0\}$. Thus, if we want to know whether the polynomial χ_A is Schur stable we perform the transformation

$$\psi_A(w) := (w-1)^d \cdot \chi_A\left(\frac{w+1}{w-1}\right) \qquad (1 \neq w \in \mathbb{C}). \tag{63}$$

It is clear that ψ_A is also a polynomial of degree d and χ_A is Schur stable if and only if ψ_A is Hurwitz stable. It is not difficult to calculate (cf. [2]) that in case of

- $d = 2$ the polynomial χ_A is Schur stable if and only if

$$-1 + |a_1| < a_0 < 1; \tag{64}$$

- $d = 3$ the polynomial χ_A is Schur stable if and only if the inequalities

$$1 + a_1 > |a_0 + a_2|, \quad 3 - a_1 > |3a_0 - a_2|, \quad 1 - a_1 > a_0(a_0 + a_2) \tag{65}$$

hold.

The second way is the application of the so called Jury test which proof is based on the Rouché theorem (cf. [16]).

2.2 Hopf bifurcation

In what follows we shall examine the situation when system (1) with $\tau := 0$ exhibits Hopf bifurcation. In order to have this, we rewrite the version of (1) without delay in the parameter-dependent form

$$\dot{x} = f \circ (x, p) \tag{66}$$

where p represents a parameter of the given system. Hopf bifurcation occurs if and only if for the eigenvalues $\mu(p) \pm i\nu(p)$ of the Jacobi matrix A of f at the critical value p_*

- the eigenvalue crossing condition holds:

$$\mu(p_*) = 0, \quad \nu(p_*) \neq 0, \quad (\sigma(A)\backslash\{\pm\nu(p_*))\}) \cap i\mathbb{R} = \emptyset; \tag{67}$$

- the transversality condition $\mu'(p_*) \neq 0$ is fulfilled.

In the case of the two-dimensional system there is a result about the fulfillment of the above two conditions.

Lemma 1. Let $I \subset \mathbb{R}$ be an open interval and $\beta, \gamma : I \to \mathbb{R}$ smooth functions. The roots of the characteristic polynomial

$$\chi_A(z) := z^2 + \beta z + \gamma \qquad (z \in \mathbb{C}) \tag{68}$$

fulfill the eigenvalue crossing condition and the transversality condition if and only if at the critical value $p = p_* \in I$

$$\beta(p_*) = 0, \qquad \gamma(p_*) > 0 \qquad \text{and} \qquad \beta'(p_*) \neq 0 \tag{69}$$

hold.

Proof:

Step 1. The polynomial χ_A has purely imaginary zeros $\pm\omega i$ with $\omega \neq 0$ if and only if

$$z^2 + \beta z + \gamma = (z - \omega i)(z + \omega i) = z^2 + \omega^2 \qquad (z \in \mathbb{C}). \tag{70}$$

Thus, at the critical value $p = p_* \in I$ the eigenvalue crossing condition holds exactly in case

$$\beta(p_*) = 0 \quad \text{and} \quad \gamma(p_*) > 0. \tag{71}$$

Step 2. Let denote by ρ the root of the equation

$$z^2 + \beta z + \gamma = 0 \tag{72}$$

which at p_* takes the value ωi: $\rho(p_*) = \omega i$, and let us introduce the following function

$$\mathcal{F}(z, p) :\equiv z^2 + \beta(p)z + \gamma(p). \tag{73}$$

Since

$$\mathcal{F}(\rho(p_*), p_*) = 0 \quad \text{and} \quad \partial_1\mathcal{F}(\rho(p_*), p_*) = 2\omega i + \beta(p_*) = 2\omega i \neq 0, \tag{74}$$

therefore we have

$$\begin{aligned} \rho'(p_*) &= -\frac{\partial_2\mathcal{F}(\omega i, p_*)}{\partial_1\mathcal{F}(\omega i, p_*)} = -\frac{\beta'(h)z + \gamma'(h)}{2z + \beta(h)}\Bigg|_{\substack{z = \omega i \\ h = p_*}} \\ &= -\frac{\beta'(p_*)\omega i + \gamma'(p_*)}{2\omega i + \beta(p_*)} =: \frac{A + Bi}{C + Di}. \end{aligned} \tag{75}$$

Using the well known calculation

$$\frac{A+Bi}{C+Di}=\frac{A+Bi}{C+Di}\cdot\frac{C-Di}{C-Di}=\frac{(AC+BD)-(AD-BC)i}{C^2+D^2}=\frac{AC+BD}{C^2+D^2}+\frac{BC-AD}{C^2+D^2}i, \tag{76}$$

furthermore the first and third part of (69) the formula

$$\frac{\mathrm{d}\Re(\rho(p_*))}{\mathrm{d}h}=\Re(\rho'(p_*))=\frac{2\omega^2\beta'(p_*)+\gamma'(p_*)\beta(p_*)}{4\omega^2+(\beta(p_*))^2}=-\frac{\beta'(p_*)}{2}\neq 0 \tag{77}$$

proves the lemma. ■

Because a matrix of order two can have no other eigenvalues besides the critical eigenvalues the crossing can happen only if for suitable $r>0$

$$\beta^2(h)-4\gamma(h)<0 \qquad (h\in(p_*-r,p_*+r)) \tag{78}$$

holds.

We have to remark that there are two forms of Hopf bifurcation: the standard one and the so called non-standard. Under standard Hopf bifurcation we mean the phenomenon when the critical eigenvalues of the Jacobian matrix A cross the imaginary axis from left to right and all other eigenvalues remain in the open left complex plane, whereas non-standard Hopf bifurcation means that the critical eigenvalues cross the imaginary axis from the right to the left and there is no restriction for the location of the other eigenvalues.

Theorem 1.7. If (78) holds then crossing can only happen

- From the left half-plane to the right, if the function β changes at the critical values p_* its sign from positive to negative;
- From the right half-plane to the left, if the function β changes at the critical values p_* its sign from negative to positive.

Proof: If condition (78) holds, then χ_A has a pair of conjugate roots. From elementary mathematics, we know that

- if function β changes its sign from positive to negative at the critical values p_*, then the sign of the real parts of the roots change their signs from the negative to positive;
- if function β changes at the critical values p_* its sign from negative to positive, then the sign of the real parts of the roots change their signs from the positive to negative. ■

Example 4. Let be $0<a,b\in\mathbb{R}$ and consider the activator-inhibitor system of Schnackenberg-type

$$\dot{x}=a-x+x^2y, \qquad \dot{y}=b-x^2y \tag{79}$$

(cf. [17]). System (79) has the unique equilibrium point

$$(x_*,y_*)=\left(a+b,\frac{b}{(a+b)^2}\right). \tag{80}$$

The Jacobian of (79) at (x_*,y_*) takes the form

$$A := J(x_*, y_*) = \begin{bmatrix} 2x_* y_* - 1 & x_*^2 \\ -2x_* y_* & -x_*^2 \end{bmatrix} \tag{81}$$

whose eigenvalues are the zeros of its characteristic polynomial

$$\begin{aligned} \chi_A(z) &= \det(zI_2 - A) = z^2 - \mathrm{Tr}(A)z + \det(A) = z^2 + \left(1 - 2x_* y_* + x_*^2\right)z + x_*^2 \\ &= z^2 + \frac{a - b + (a+b)^3}{a+b} z + (a+b)^2 =: z^2 + \beta z + \gamma \quad (z \in \mathbb{C}). \end{aligned} \tag{82}$$

It is easy to see that if we choose $b =: p$ as parameter by fixed a then for every $p(>0)$ we have $\gamma(p) > 0$ and

$$\beta'(p) = \frac{\left[-1 + 3(a+p)^2\right](a+p) - a + p - (a+p)^3}{(a+p)^2} = \frac{2(a+p)^3 - 2a}{(a+p)^2} \neq 0. \tag{83}$$

This means that Hopf bifurcation occurs at the critical value p_* if and only if p_* is a positive real solution of the equation

$$\kappa(p) := (a+p) \cdot \beta(h) = a - p + (a+p)^3 = 0 \tag{84}$$

extended in p to the whole real axis. For example,

- In case of $a = 1$ the polynomial κ has one real root:

$$\kappa(p) = 0 \iff p \in \left\{0, \frac{-3 - \sqrt{7}i}{2}, \frac{-3 + \sqrt{7}i}{2}\right\}. \tag{85}$$

Clearly, $\kappa(1) = 8 > 0$, therefore κ and so β assume on the positive half line positive values, which has a consequence that the characteristic polynomial and hence the equilibrium point (x_*, y_*) is stable, since a second order characteristic polynomial is stable if and only if its coefficients have the same (positive) sign.

- in case of $a = 0.1$ the polynomial κ has three real roots:

$$\kappa(h) = 0 \iff h \in \{p_1 := -1.18803\ldots,\ p_2 := 0.109149\ldots,\ p_3 := 0.778885\ldots\}. \tag{86}$$

Because

$$\kappa(0) = a + a^3 > 0, \qquad \kappa(0.5) = 0.184 < 0, \qquad \kappa(1) = a\left(a^2 + 3a + 4\right) > 0, \tag{87}$$

the polynomial κ and so β changes its sign at p_2 from positive to negative, and at p_3 from negative to positive. This means in the light of the above that at the parameter value $p = p_2$ standard Hopf bifurcation occurs: the roots migrate from the left open half plane to the right, (x_*, y_*) loses its stability; furthermore at the parameter value $p = p_3$ non-standard Hop bifurcation takes place, i.e. the roots migrate from the right half plane to the left and as a consequence (x_*, y_*) becomes stable.

In the case of the three-dimensional system we post a result (cf. [18]), the proof of which is similar to the one in [19].

Lemma 2. Let be $I \subset \mathbb{R}$ an open interval and $\alpha, \beta, \gamma : I \to \mathbb{R}$ smooth functions. The roots of the characteristic polynomial

$$\chi_A(z) := z^3 + \alpha z^2 + \beta z + \gamma \qquad (z \in \mathbb{C}) \tag{88}$$

fulfill the crossing and the transversality conditions if and only if at the critical value $p = p_* \in I$

$$\beta(p_*) > 0, \quad \alpha(p_*) \neq 0, \quad \gamma(p_*) = \alpha(p_*)\beta(p_*) \tag{89}$$

furthermore

$$\frac{\mathrm{d}}{\mathrm{d}h}\{\alpha(h)\beta(h) - \gamma(h)\}|_{h=p_*} \neq 0 \tag{90}$$

hold.

Proof:

Step 1. We show that the characteristic polynomial χ_A has purely imaginary roots $\pm\omega i$ $(\omega \neq 0)$, if $\alpha\beta - \gamma = 0$ and $\beta > 0$ hold. Indeed, if ξ, η, ζ denote the roots of the polynomial χ_A, then using Orlando formula we have

$$\beta = \xi\eta + \xi\zeta + \eta\zeta, \text{ resp. } \alpha\beta - \gamma = -(\xi + \eta)(\xi + \zeta)(\eta + \zeta). \tag{91}$$

This means that

- if for some $0 \neq \omega \in \mathbb{R}$ the equalities

$$\chi_A(\omega i) = 0 = \chi_A(-\omega i) \tag{92}$$

hold, then one of the three zeros, like ζ is real, furthermore

$$\xi = \omega i = -\eta, \text{ resp. } \beta = \omega^2 + \zeta\omega i - \zeta\omega i = \omega^2 > 0; \tag{93}$$

- χ_A has exactly zeros with opposite sign but the same absolute value, if $\alpha\beta - \gamma = 0$ holds, furthermore χ_A has a complex root, since if $\xi = -\eta$ then

$$\beta = -\xi^2 + \xi\zeta - \xi\zeta = -\xi^2 \leq 0, \tag{94}$$

which contradicts the fact that $\beta > 0$.

It is clear that from conditions (89) it follows that the zeros of χ_A are

$$-\alpha \text{ and } \pm\sqrt{\beta}i = \pm\sqrt{\frac{\gamma}{\alpha}}i, \tag{95}$$

because

$$\begin{aligned}(z+\alpha)\left(z - \sqrt{\beta}i\right)\left(z + \sqrt{\beta}i\right) &\equiv (z+\alpha)\left(z^2 + \beta\right) \equiv z^3 + \alpha z^2 + \beta z + \alpha\beta \\ &\equiv z^3 + \alpha z^2 + \beta z + \gamma.\end{aligned} \tag{96}$$

Step 2. Let denote the roots of χ_A by ρ which assumes at p_* the value ωi: $\rho(p_*) = \omega i$, furthermore let define

$$\mathcal{F}(z, h) :\equiv z^3 + \alpha(h)z^2 + \beta(h)z + \gamma(h). \tag{97}$$

Because

$$\mathcal{F}(\rho(p_*),p_*) = 0, \text{ and } \partial_1\mathcal{F}(\omega i,p_*) = \beta(p_*) - 3\omega^2 + 2\alpha(p_*)\omega i \neq 0, \tag{98}$$

it follows for the derivative of the implicit function ρ at p_* that

$$\begin{aligned} \rho'(p_*) &= -\frac{\partial_2\mathcal{F}(\omega i,p_*)}{\partial_1\mathcal{F}(\omega i,p_*)} = -\frac{\alpha'(h)z^2 + \beta'(h)z + \gamma'(h)}{3z^2 + 2\alpha(h)z + \beta(h)}\Bigg|_{\substack{z=\omega i\\ h=p_*}} \\ &= \frac{\alpha'(p_*)\omega^2 - \beta'(p_*)\omega i - \gamma'(p_*)}{-3\omega^2 + 2\alpha(p_*)\omega i + \beta(p_*)}. \end{aligned} \tag{99}$$

Using (89), resp. (90), furthermore $\beta(p_*) = \omega^2$ we have

$$\begin{aligned} \frac{d\Re(\rho(p_*))}{dh} &= \Re(\rho'(p_*)) \\ &= \frac{\{\alpha'(p_*)\omega^2 - \gamma'(p_*)\}\{\beta(p_*) - 3\omega^2\} - 2\alpha(p_*)\beta'(p_*)\omega^2}{[\beta(p_*) - 3\omega^2]^2 + 4\alpha(p_*)^2\omega^2} \\ &= \frac{\{\alpha'(p_*)\beta(p_*) - \gamma'(p_*)\}\{-2\beta(p_*)\} - 2\alpha(p_*)\beta'(p_*)\beta(p_*)}{[-2\beta(p_*)]^2 + 4\alpha(p_*)^2\beta(p_*)} \\ &= \frac{2\beta(p_*)\cdot\{\gamma'(p_*) - \alpha'(p_*)\beta(p_*) - \alpha(p_*)\beta'(p_*)\}}{4\beta(p_*)\left\{\beta(p_*) + \alpha(p_*)^2\right\}} = \\ &= \frac{(\gamma - \alpha\beta)'(p_*)}{2\left\{\beta(p_*) + \alpha(p_*)^2\right\}} \neq 0. \quad \blacksquare \end{aligned} \tag{100}$$

As a consequence of the Stodola criterion the necessary condition for the stability of the polynomial χ_A is the positivity of its coefficients α, β, γ. In the case when its zeros cross the imaginary axis from the right to the left, one of the coefficients should be local negative.

Theorem 1.8. If conditions of Lemma 2., i.e. (89) and (90) hold, furthermore, for every $\varepsilon > 0$, resp. arbitrary $p_* \neq h \in (p_* - \varepsilon, p_* + \varepsilon)$ the equalities

$$\beta(h) = \alpha^2(h) \text{ and } \gamma(h) \notin \{0, \zeta(h)\} \tag{101}$$

are valid, where

$$\zeta(h) := \frac{3\alpha(h)\beta(h) - 2\alpha^2(h)}{9} \quad (h \in I), \tag{102}$$

then the roots of χ_A cross the imaginary axis

- from left to right, if

$$3\sqrt[3]{\gamma(h) - \zeta(h)} < 2\alpha(h) \tag{103}$$

- from right to left, if

$$3\sqrt[3]{\gamma(h) - \zeta(h)} > 2\alpha(h) \tag{104}$$

hold.

Proof: Using the notations

$$a := \frac{\alpha}{3}, \quad b := \frac{\beta}{3}, \quad \text{resp.} \quad A := a^2 - b, \quad B := 2a^2 - 3ab + \gamma \tag{105}$$

one can see that if $A = 0$ and $0 \neq B \in \mathbb{R}$ hold then the zeros of the polynomial χ_A are as follows (cf. [17]):

$$\xi = -\sqrt[3]{B} - a, \qquad \eta = \frac{\sqrt[3]{B}}{2} - a + \frac{3\sqrt[3]{B^2}}{4}i, \qquad \zeta = \frac{\sqrt[3]{B}}{2} - a + \frac{3\sqrt[3]{B^2}}{4}i. \tag{106}$$

Thus, from $B \neq 0$ and $\sqrt[3]{B} \neq 2a$ it follows that χ_A has a pair of complex conjugate roots. Because of condition $\gamma \neq 0$ the third root could not be zero, furthermore this pair of complex conjugate zeros lies

- in the left half-plane if and only if $\sqrt[3]{B} < 2a$, i.e. (103) holds;
- in the right half-plane if and only if $\sqrt[3]{B} > 2a$, i.e. (104) holds. ■

Example 5. In [20] Liao, Zhou, and Tang proposed the following autonomous system of ordinary differential equations

$$\dot{x} = a(y - x), \qquad \dot{y} = dx + cy - xz, \qquad \dot{z} = -bz + xy \tag{107}$$

where $a, b, c, d \in \mathbb{R}$. If (a, b, c, d) belongs to the set

$$\{(a, b, -1, d), \quad (a, b, c, c - a), \quad (a, b, c, 0)\} \tag{108}$$

then we get the Lorenz system (cf. [21]), the Chen system (cf. [22]) and the Lü system (cf. [23]). Yan showed (cf. [24]) that in the system (107) Hopf bifurcation may occur. In what follows we show that his calculations can be simplified as we know the coefficients of the characteristic polynomials of the Jacobian of system (107). If $b(c + d) > 0$ holds then system (107) has three equilibria:

$$E_0 := (0, 0, 0), \qquad E_\pm := \left(\pm\sqrt{b(c+d)}, \pm\sqrt{b(c+d)}, c + d\right). \tag{109}$$

The Jacobian of (107) takes the form

$$J(x, y, z) := \begin{bmatrix} -a & a & 0 \\ d - z & c & -x \\ y & x & -b \end{bmatrix} \qquad ((x, y, z) \in \mathbb{R}^3). \tag{110}$$

Hence the corresponding Jacobians are:

$$A_0 := J(E_0) := \begin{bmatrix} -a & a & 0 \\ d & c & 0 \\ 0 & 0 & -b \end{bmatrix}, \tag{111}$$

resp.

$$A_\pm := J(E_\pm) := \begin{bmatrix} -a & a & 0 \\ d - (c + d) & c & \mp\sqrt{b(c+d)} \\ \pm\sqrt{b(c+d)} & \pm\sqrt{b(c+d)} & -b \end{bmatrix}. \tag{112}$$

The eigenvalues of A_0 are the roots of the characteristic polynomial

$$\chi_{A_0}(\xi) := \det(\xi I_3 - J_0) = \xi^3 + \alpha\xi^2 + \beta\xi + \gamma \qquad (\xi \in \mathbb{C}), \tag{113}$$

where $\gamma := -\det(A_0) = -ab(c+d)$, $\beta := \mathrm{Tr}(\mathrm{adj}(A_0)) = ab - bc - ac - ad$ and $\alpha := -\mathrm{Tr}(A_0) = a + b - c$.

We remark that only a special parameter configuration was investigated in [24], namely $a = c, d = -2c$ $(c > 0)$. On the other hand one can observe more due to Lemma 2:

1. it is easy to see from the characteristic polynomial that in case $c = a + b$ Hopf bifurcation may not occur because no matter what will be chosen as a bifurcation parameter, the coefficient of the first-order term of the polynomial vanishes, which has the consequence (cf. Lemma 2) that there can be a pair of complex conjugate zeros on the imaginary axis only in case if the constant term of the polynomial vanishes. This contradicts the crossing condition.

2. the parameter d cannot be considered as a bifurcation parameter, because as its value is changed Hopf bifurcation may not occur. This can be explained as follows. If $a \neq 0$ then

$$(\gamma - \alpha\beta)(d_*) = 0 \quad \Longleftrightarrow \quad d_* = \frac{ab + b^2 - ac - bc}{a}$$

(in case of $a = 0$ the system would be two dimensional) and $\beta(d_*) = -b^2$, this contradicts the first condition in (89).

3. the parameter c can be chosen as a bifurcation parameter only under some restrictions, because in the case of $a \neq -b$

$$(\gamma - \alpha\beta)(c_*) = 0 \quad \Longleftrightarrow \quad c_* \in \left\{ a, \ \frac{ab + b^2 - ad}{a + b} \right\}$$

and

$$\beta(c_*) \in \{-a(a+d), \ -b^2\}$$

will be positive only in case if a and $a + d$ have opposite signs as it is in case $a = c, d = -2c$ $(c > 0)$ proposed in [24]. If $a = -b$ and $a \neq 0$ then

$$(\gamma - \alpha\beta)(c_*) = 0 \quad \Longleftrightarrow \quad c_* = \frac{ab + b^2 - ac - bc}{a}$$

(in case of $a = 0$ the system would be two dimensional) and $\beta(c_*) = -b^2$ which contradicts the first condition in (89). If $a = c$ then

$$(\gamma - \alpha\beta)(b_*) = 0 \quad \Longleftrightarrow \quad d = 0.$$

In this case we have

$$\frac{\mathrm{d}}{\mathrm{d}b}(\gamma - \alpha\beta)(b_*) = 0,$$

which contradicts the transversality condition.

It is easy to see the following: by fixing the parameters b, c, resp. $d = -2c$ $(c > 0)$ (cf. [24]) the parameter a will be chosen as bifurcation parameter then at value $a_* := c$ Hopf bifurcation takes place, because with substitution $d = -2c$ we have

$$\beta(a_*) = cb - bc - cc - c(-2c) = c^2 > 0, \qquad \alpha(a_*) = b \neq 0, \tag{114}$$

resp.

$$(\gamma - \alpha\beta)(a_*) = [abc - (a + b - c)(ab - bc + ac)]_{a=c} = bc^2 - bc^2 = 0 \tag{115}$$

and

$$\frac{\mathrm{d}}{\mathrm{d}a}\{\gamma - \alpha\beta\}(a_*) = \frac{\mathrm{d}}{\mathrm{d}a}(abc - (a + b - c)(ab - bc + ac))\Big|_{a=c}$$

$$= [bc - (ab - bc + ac) - (a + b - c)(b + c)]_{a=c} = bc - c^2 - b(b + c) = -c^2 - b^2 \neq 0. \tag{116}$$

The eigenvalues of the matrix $A_\pm$ are the zeros of the characteristic polynomial

$$\chi_{A_\pm}(\xi) := \det(\xi I_3 - A_\pm) = \xi^3 + \alpha\xi^2 + \beta\xi + \gamma \qquad (\xi \in \mathbb{C}) \tag{117}$$

where

$$\gamma := -\det(J) = 2ab(c + d), \quad \beta := \mathrm{Tr}(\mathrm{adj}(J)) = b(a + d), \quad \alpha := -\mathrm{Tr}(J) = a + b - c. \tag{118}$$

If

$$a(3a + d)(a + b + d) \neq 0, \tag{119}$$

then at

$$c_* = \frac{a^2 + ab - ad + bd}{3a + d} \tag{120}$$

Hopf bifurcation takes place, because

$$\beta(c_*) = b(a + d) > 0, \qquad \alpha(c_*) = \frac{2a(a + b + d)}{3a + d}, \qquad (\gamma - \alpha\beta)(c_*) = 0, \tag{121}$$

resp.

$$\frac{\mathrm{d}}{\mathrm{d}c}(\gamma - \alpha\beta)(c_*) = -2ab - b(a + d) = -b(3a + d) \neq 0. \tag{122}$$

3. The delayed case: $\tau > 0$

When modeling and analyzing processes and behaviors which come from a natural environment, it often happens that we need a bit of distance in time to see the changes of the considered quantities (which are the variables of our model). For example, when we think about the epidemiological models, it is a well-founded thought that we need some time while susceptibles become infectious, and hence it is reasonable to assume that the migration of the individuals from the class of susceptibles into the infected is subject to delay.

Another expressive example is the modeling of the processes of the human body or the brain, like emotions: love, hate, etc. A bit of time has to pass for the brain to process the signals coming from various places, and only after this delay, the mood could change. These changes can be described and analyzed with delayed differential equations. One type of these systems is the so called Romeo and Juliet model, where the changes of Romeo's and Juliet's love and hate in time are described as a system of two linear ordinary differential equations. In this chapter we are going to consider this model with general coefficients and investigate the stability of the linear system.

We are going to consider the following linear system:

$$\begin{aligned} \dot{x} &= \alpha_1 x + A_1 x(\cdot - \tau) + \alpha_2 y + A_2 y(\cdot - \tau), \\ \dot{y} &= \alpha_3 x + A_3 x(\cdot - \tau) + \alpha_4 y + A_4 y(\cdot - \tau), \end{aligned} \tag{123}$$

where A_i, $\alpha_i \in \mathbb{R}$, $(i \in \{1, 2, 3, 4\})$, $\tau > 0$, with initial conditions $\Phi = (\Phi_1, \Phi_2)$ in the Banach space

$$\{\Phi \in C([-\tau, 0], \mathbb{R}^2_+) : \ \Phi_1(\theta) = x(\theta), \Phi_2(\theta) = y(\theta)\}, \tag{124}$$

where $\Phi_i(\theta) > 0$, $(\theta \in [-\tau, 0], i \in \{1, 2\})$. Straightforward calculation shows that the characteristic function of the above system (with regard to the trivial equilibrium point) takes the form:

$$\begin{aligned} \Delta(z, \tau) \equiv z^2 - (\alpha_1 + \alpha_4)z + \alpha_1\alpha_4 - \alpha_2\alpha_3 + {+e^{-z\tau}} \cdot (-(A_1 + A_4)z + \alpha_1 A_4 \\ +\alpha_4 A_1 - \alpha_2 A_3 - \alpha_3 A_2) + e^{-2z\tau}(A_1 A_4 - A_2 A_3) \quad (z \in \mathbb{C}). \end{aligned} \tag{125}$$

In [25] the authors treat delay differential equations, which characteristic function for arbitrary $z \in \mathbb{C}$ has the form

$$\Delta(z, \tau) = z^2 + a_1 z + a_0 + (b_1 z + b_0)e^{-z\tau} + ce^{-2z\tau} \tag{126}$$

where a_0, a_1, b_0, b_1, c are arbitrary real constants. It can be seen that the characteristic functions (125) and (126) has the same form. In [25] the authors assume that $c = 0$ to simplify the analysis. Furthermore, they say that due to the continuous dependence of eigenvalues of the model parameters (cf. [26]) their results are valid for sufficiently small c parameters, too. Nevertheless, in the literature a lot of models and systems are investigated in which the coefficient of $e^{-2z\tau}$ of the characteristic function is not equal to zero, and maybe not sufficiently small. Therefore, the aim of this section is to show that we can analyze the stability of the system in the case where $c \neq 0$, too.

In this section we assume that $\tau > 0$ holds and investigate the qualitative behavior of the linearized system (3), more precisely we study the stability of characteristic function

$$\Delta(z, \tau) :\equiv p(z) + q(z)e^{-z\tau} + r(z)e^{-2z\tau} \tag{127}$$

where p, q and r are polynomials with real coefficients and $\deg(r) < \deg(q)$. Under stability of Δ we mean that the zeros of Δ lie in the open left half of the complex plane. Using the Mikhailov criterion we give for special p, q and r fulfilling the above condition an estimate on the length of delay τ for which no stability switching occurs. Then for special parameters we compare our results with other methods. It follows then a delay independent stability analysis. Finally, a formula for Hopf bifurcation is calculated in terms of p, q and r. If we assume that the characteristic function has the form as in (126), then we can give conditions easily on the parameters a_1, a_0, b_1, b_0, c and an upper bound τ_1 such that with $\tau < \tau_1$ the system is asymptotically stable. In other words stability change may happen only for $\tau \geq \tau_1$.

In what follows, the Mikhailov stability criterion will be proved, which is the implication of the argument principle (cf. [27, 28]). The treatment is based on [29].

Theorem 1.9 (Mikhailov criterion). Consider the quasi-polynomial

$$M(z) := Q(z) + \sum_{k=1}^{p} R_k(z) e^{-s\tau_k} \qquad (z \in \mathbb{C}), \tag{128}$$

where the order of the polynomials Q and R_k is less than or equal to $d \in \mathbb{N}$, and they are defined as

$$Q(z) := q_d z^d + \dots + q_0, \quad R_k(z) := r_{k_d} z^d + \dots + r_{k_0} \qquad (z \in \mathbb{C}) \tag{129}$$

where $q_i, r_{k_i} \in \mathbb{R}$ for $i = 1, \dots, d$, $k = 1, \dots, p$, $q_d > 0$ and

$$\max_{k \in \{1, \dots, p\}} (\deg(R_k)) < d, \tag{130}$$

furthermore $\tau_k \geq 0$ for $k = 1, \dots, p$. If M defined by (128) has no zeros on the imaginary axis, then M is stable if and only if

$$\Delta := \Delta_{\omega \in [0, +\infty)} \arg(M(i\omega)) = \frac{d\pi}{2} \tag{131}$$

holds where Δ denotes the change of argument of the vector $M(i\omega)$ anticlockwise in the complex plane as ω increases from 0 to $+\infty$.

Proof: In order to prove the theorem, we will apply the argument principle (cf. [28]) to M on the Γ-contour (cf. **Figure 2**) where $\Gamma =: C_1 \cup C_2$ denotes the

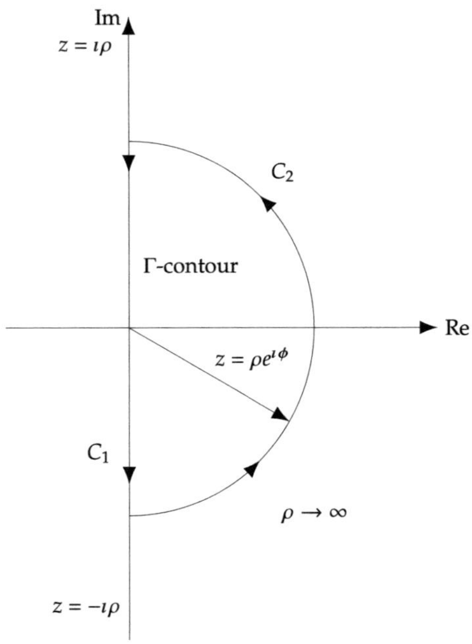

Figure 2.
Γ-contour on the complex plane.

positive oriented curve in the complex plane which consists of the interval $[-\rho, \rho]$ $(\rho > 0)$ on the imaginary axis, i.e.

$$C_1 := \{is \in \mathbb{C} : \ s \in [-\rho, \rho]\} \tag{132}$$

and the semicircle C_2 of the radius ρ in the right-hand half-plane:

$$C_2 := \left\{\rho e^{i\phi} \in \mathbb{C} : \ \phi \in \left[-\frac{\pi}{2}, \frac{\pi}{2}\right]\right\}. \tag{133}$$

Since

$$\max_{k \in \{1, \dots, p\}} (\deg(R_k)) < \deg(Q), \tag{134}$$

there is only a finite number of roots of M in the right-half plane.

On C_2 the characteristic equation can be written for every $z \in \mathbb{C}$ as follows:

$$M(z) = q_d\rho^d(\cos(d\phi) + i\sin(d\phi)) + \dots + q_0$$
$$+ \sum_{k=1}^{p}\left\{r_{k(d-1)}\rho^{d-1}(\cos((d-1)\phi) + i\sin((d-1)\phi))e^{-\rho\tau_k e^{i\phi}} + \dots + r_{k_0}e^{-\rho\tau_k e^{i\phi}}\right\}. \tag{135}$$

Now from the summation, we can write a typical term as

$$\begin{aligned} r_{k(d-1)}\rho^{d-1}e^{i(d-1)\phi}e^{-\rho\tau_k e^{i\phi}} &= r_{k(d-1)}\rho^{d-1}e^{i(d-1)\phi}e^{-\rho\tau_k\cos(\phi) - i\rho\tau_k\sin(\phi)} \\ &= r_{k(d-1)}\rho^{d-1}e^{-\rho\tau_k\cos(\phi)}e^{i((d-1)\phi - \rho\tau_k\sin(\phi))} \end{aligned} \tag{136}$$

Therefore,

$$M(z) = q_d\rho^d(\cos(d\phi) + i\sin(d\phi)) + \dots + q_0 + \sum_{k=1}^{p} e^{-\rho\tau_k\cos(\phi)}$$
$$\cdot\left\{r_{k(d-1)}\rho^{d-1}[\cos((d-1)\phi - \rho\tau_k\sin(\phi)) + i\sin((d-1)\phi - \rho\tau_k\sin(\phi))]\right.$$
$$+ r_{k(d-2)}\rho^{d-2}[\cos((d-2)\phi - \rho\tau_k\sin(\phi)) + i\sin((d-2)\phi - \rho\tau_k\sin(\phi))]$$
$$\left. + \dots + r_{k_0}[\cos(\rho\tau_k\sin(\phi)) - i\sin(\rho\tau_k\sin(\phi))]\right\}. \tag{137}$$

Hence the argument or phase θ of the vector $M(z)$ on C_2 may be written

$$\tan(\theta) = \frac{\sin(\theta)}{\cos(\theta)} =: \frac{A}{B}, \tag{138}$$

where

$$\begin{aligned} A := \sin(\theta) = q_d\rho^d\sin(d\phi) + \dots + q_1\rho\sin(\phi)\sum_{k=1}^{p} e^{-\rho\tau_k\cos(\phi)} \\ \cdot\{r_{k(d-1)}\rho^{d-1}\sin((d-1)\phi - \rho\tau_k\sin(\phi)) \\ + \dots + r_{k_1}\rho\sin(\phi - \rho\tau_k\sin(\phi)) - r_{k_0}\sin(\rho\tau_k\sin(\phi))\} \end{aligned} \tag{139}$$

and

$$B := \cos(\theta) = q_d\rho^d \cos(d\phi) + \ldots + q_1\rho\cos(\phi) + q_0 + \sum_{k=1}^{p} e^{-\rho\tau_k \cos(\phi)} \cdot \left\{ r_{k(d-1)}\rho^{d-1}\cos((d-1)\phi - \rho\tau_k\sin(\phi)) + \ldots + r_{k_0}\cos(\rho\tau_k\sin(\phi)) \right\}. \tag{140}$$

Dividing the numerator and denominator by ρ^d gives

$$\tan(\theta) =: \frac{A_1}{B_1}, \tag{141}$$

where

$$A_1 := q_d \sin(d\phi) + \ldots + \frac{q_1}{\rho^{d-1}}\sin(\phi) + \sum_{k=1}^{p} e^{-\rho\tau_k\cos(\phi)} \cdot \left\{ \frac{r_{k(d-1)}}{\rho}\sin((d-1)\phi - \rho\tau_k\sin(\phi)) + \ldots + \frac{r_{k_1}}{\rho^{d-1}}\sin(\phi - \rho\tau_k\sin(\phi)) - \frac{r_{k_0}}{\rho^d}\sin(\rho\tau_k\sin(\phi)) \right\} \tag{142}$$

and

$$B_1 := q_d\cos(d\phi) + \ldots + \frac{q_1}{\rho^{d-1}}\cos(\phi) + \frac{q_0}{\rho^d} + \sum_{k=1}^{p} e^{-\rho\tau_k\cos(\phi)} \cdot \left\{ \frac{r_{k(d-1)}}{\rho}\cos((d-1)\phi - \rho\tau_k\sin(\phi)) + \ldots + \frac{r_{k_0}}{\rho^d}\cos(\rho\tau_k\sin(\phi)) \right\}. \tag{143}$$

Now since

$$|\cos(\alpha)| \leq 1 \text{ and } |\sin(\alpha)| \leq 1 \qquad (\alpha \in \mathbb{R}) \tag{144}$$

and since

$$\cos(\alpha) \geq 0 \qquad \left(\alpha \in \left[-\frac{\pi}{2}, \frac{\pi}{2}\right]\right), \tag{145}$$

we have

$$\tan(\theta) = \frac{\sin(d\phi)}{\cos(n\phi)} = \tan(d\phi) \tag{146}$$

or

$$\theta = d\phi + m\pi, \qquad m \in \mathbb{Z} \tag{147}$$

as $\rho \to \infty$. Therefore, the change in argument of $M(z)$ on C_2 is given by

$$\Delta_{C_2}\arg(M) = \frac{d\pi}{2} + m\pi - \left(-\frac{d\pi}{2} + m\pi\right) = d\pi. \tag{148}$$

Now from the argument principle we can write

$$\Delta_{C_1}\arg(M) + \Delta_{C_2}\arg(M) = 2\pi N, \tag{149}$$

where N is the total number of zeros of M inside Γ. Therefore

$$\Delta_{C_1}\arg(M) = 2\pi N - d\pi. \tag{150}$$

If we reverse the direction of integration along C_1 and note the symmetry about the real axis, we have

$$\Delta_{\omega\in[0,+\infty)}\arg(M(i\omega)) = \frac{1}{2}(d\pi - 2N) = \frac{d\pi}{2} - \pi N. \tag{151}$$

in case of stability we have $N = 0$. Hence, stability requires

$$\Delta_{\omega\in[0,+\infty)}\arg(M(i\omega)) = \frac{d\pi}{2}. \quad \blacksquare \tag{152}$$

As a consequence, we have the following.

Lemma 3. Let p, q and r be polynomials, with condition

$$\deg(p) > \max\{\deg(q), \deg(r)\}, \tag{153}$$

and assume that the quasi-polynomial in (127) has no roots on the imaginary axis. Then $\Delta(\cdot, \tau)$ is stable, i.e. all of its roots have negative real part if and only if

$$[arg\Delta(i\omega, \tau)]_{\omega=0}^{\omega=+\infty} = \frac{\pi}{2}\cdot\deg(p(i\omega)), \tag{154}$$

i.e. the argument of $\Delta(i\omega, \tau)$ increases $\pi/2\cdot\deg(p(i\omega))$ as ω increases from 0 to $+\infty$. □

Theorem 1.10 If for the delay parameter τ in the characteristic function (126)

$$\tau < \frac{a_1 - |b_1|}{|b_0| + 2|c|}, \qquad a_0 + b_0 + c > 0 \tag{155}$$

hold, then the characteristic function, and hence the trivial equilibrium point of system (123) is asymptotically stable.

Proof: Substituting $z = i\omega$ $(\omega > 0)$ into (126), we get

$$\Delta(i\omega, \tau) = p(i\omega) + q(i\omega)e^{-i\omega\tau} + r(i\omega)e^{-2i\omega\tau}. \tag{156}$$

Hence using the characteristic function (126), where

$$p(z) \equiv z^2 + a_1 z + a_0, \quad q(z) \equiv b_1 z + b_0, \quad r(z) \equiv c \tag{157}$$

we have for $\omega > 0$ that

$$\Delta_R(i\omega, \tau) \equiv -\omega^2 + a_0 + b_0\cos(\omega\tau) + b_1\omega\sin(\omega\tau) + c\cos^2(\omega\tau) - c\sin^2(\omega\tau), \tag{158}$$

and also

$$\Delta_I(i\omega, \tau) \equiv a_1\omega + b_1\omega\cos(\omega\tau) - b_0\sin(\omega\tau) - 2c\sin(\omega\tau)\cos(\omega\tau). \tag{159}$$

It could be seen that

$$\Delta_R(0,\tau) = a_0 + b_0 + c > 0 \quad \text{and} \quad \Delta_I(0,\tau) = 0, \tag{160}$$

furthermore

$$\lim_{\omega \to +\infty} \Delta_R(i\omega,\tau) = -\infty. \tag{161}$$

Therefore, we have to show that $\Delta_I(i\omega,\tau) > 0$ for each $\omega > 0$. If it holds, then

$$arg(\Delta(i\omega,\tau)) = \pi, \tag{162}$$

and hence by using the Mikhailov criterion we have stability.

Substituting $w := \omega\tau$ into (159) and multiplying the result by τ, we get

$$\tau\Delta_I\left(i\frac{w}{\tau},\tau\right) = a_1 w + b_1 w \cos(w) - \tau b_0 \sin(w) - 2\tau c \sin(w)\cos(w). \tag{163}$$

By using straightforward estimations we obtain that

$$\tau\Delta_I\left(i\frac{\omega}{\tau},\tau\right) > (a_1 - |b_1| - \tau(|b_0| + 2|c|))\omega, \tag{164}$$

and hence $\Delta_I(i\omega,\tau) > 0$ fulfills for $\omega > 0$, if the first condition of (155) is satisfied, i.e. if

$$\tau < \frac{a_1 - |b_1|}{|b_0| + 2|c|} \tag{165}$$

fulfills. ■

We show now a simple example in order to demonstrate the above theorem and what the conditions say. First of all, we are going to see that the conditions of the previous theorem are sufficient, but not necessary.

Example 6. Let us consider the following system of two linear delay differential equations.

$$\dot{x} = -\frac{1+\sqrt{7}}{2}x + x(\cdot - \tau) + \frac{-4+\sqrt{7}}{2}y - \frac{1}{2}y(\cdot - \tau), \quad \dot{y} = x + x(\cdot - \tau) + \frac{-3+\sqrt{7}}{2}y. \tag{166}$$

The characteristic polynomial of (166) is

$$\Delta(z,\tau) = z^2 + 2z + 1 + (z-1)e^{-z\tau} + \frac{1}{2}e^{-2z\tau} \qquad (z \in \mathbb{C}, \tau \geq 0). \tag{167}$$

Since $a_0 + b_0 + c = \frac{1}{2} > 0$, i.e. the second condition in Theorem 1.10 fulfills, we know from that theorem that if

$$\tau < \frac{a_1 - |b_1|}{|b_0| + 2|c|} = \frac{1}{2}, \tag{168}$$

then the trivial solution of (166) is asymptotically stable. As earlier mentioned, the conditions of Theorem 1.10 are sufficient, but not necessary, which can be easily seen, if we study the phase portrait of the system (166) with the following different values of the parameter τ: firstly with $\tau = 0.48$, then with $\tau = 0.6$ and finally with $\tau = 0.94$. The **Figures 3** and **4** represent the solutions of system (166) with different values of the parameter τ.

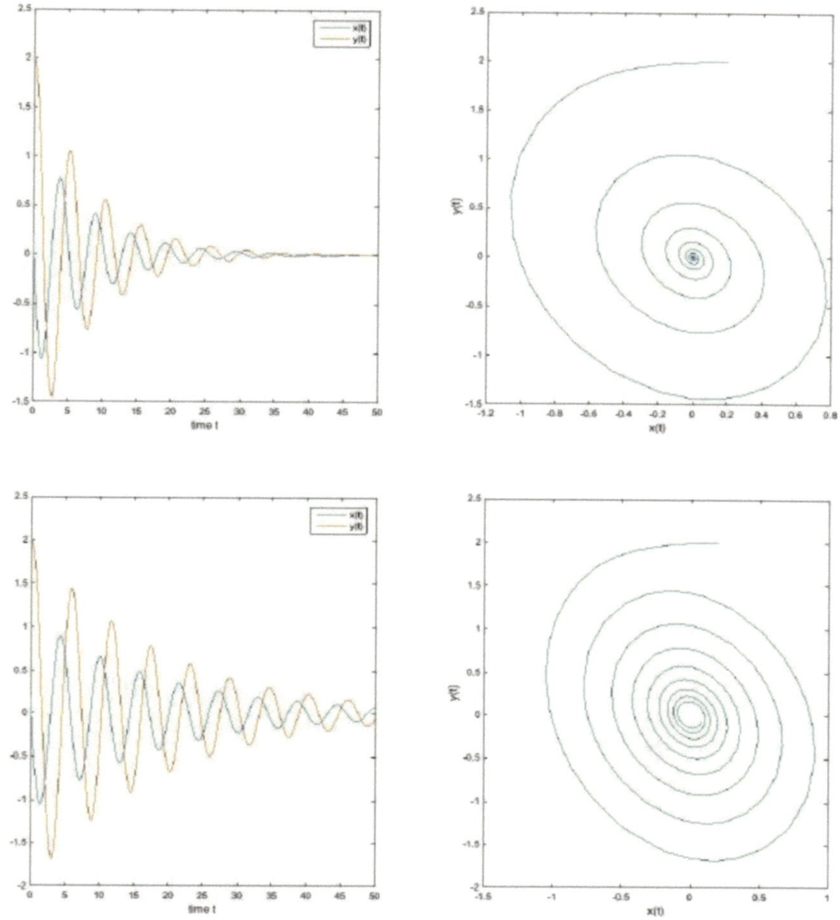

Figure 3.
Example: solutions with $\tau = 0.48$ *and* $\tau = 0.6$.

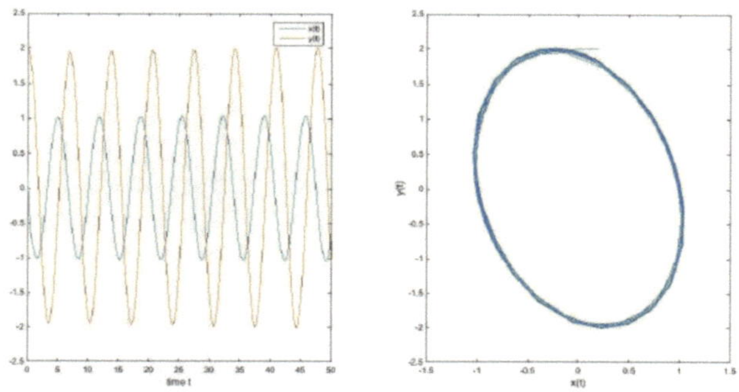

Figure 4.
Example: solutions with $\tau = 0.94$.

In **Figure 3** (above) the parameter τ is less than half, so the parameters fulfill the condition of Theorem 1.10, and the origin is asymptotically stable. In **Figure 3** (bottom) the parameter value shows that the theorem does not give a necessary condition, because here the value of the parameter is bigger than half, but the origin is still asymptotically stable. But if we increase more the value of the parameter τ, the quasi-polynomial and hence the origin changes to unstable.

The previous example shows that it would be useful to give the largest bound in the theorem, because if we have a larger bound, then we can guarantee the stability of the quasi-polynomial for higher value of the parameter τ. In this sense, we can compare our result in Theorem 1.10 with another result in the literature. In this chapter, we compare the conditions of the theorem coming from [30, 31]. In [31] Stépán considered the system

$$\dot{x} = -a_{11}x - a_{12}y + b_{11}x(\cdot - \tau), \qquad \dot{y} = -a_{21}x - a_{22}yb_{22}y(\cdot - \tau) \tag{169}$$

where $a_{11}, a_{12}, a_{21}, a_{22}, b_{11}, b_{22} > 0$ and $\tau \geq 0$. The characteristic function of system (169) is the quasi-polynomial

$$\begin{aligned}\Delta_S(z, \tau) :=& z^2 + (a_{11} + a_{22})z + a_{11}a_{22} - a_{12}a_{21} - ((b_{11} + b_{22})z + a_{22}b_{11} \\ &+ a_{11}b_{22}) \cdot e^{-z\tau} + b_{11}b_{22}e^{-2z\tau} \qquad (z \in \mathbb{C}, \tau \geq 0).\end{aligned} \tag{170}$$

In [30] Freedman and Rao worked with the system

$$\dot{x} = -D_1x - D_2y + B_1x(\cdot - \tau), \qquad \dot{y} = -F_1x - F_2y + E_2y(\cdot - \tau) \tag{171}$$

where $B_1, D_1, D_2, E_2, F_1, F_2 > 0$ are constants and the characteristic function of (171) is the quasi-polynomial

$$\begin{aligned}\Delta_{FR}(z, \tau) :=& z^2 + (D_1 + F_2)z + D_1F_2 - D_2F_1 - ((B_1 + E_2)z + B_1F_2 \\ &+ D_1E_2) \cdot e^{-z\tau} + B_1E_2e^{-2z\tau} \qquad (z \in \mathbb{C}, \tau \geq 0).\end{aligned} \tag{172}$$

Similarly to papers [30, 31] we gave an upper bound for τ under which the quasi-polynomial is Hurwitz stable.

Let us write these conditions for τ using the notations of Stépán. In [31] we can find the condition

$$\tau < \frac{a_{11} + a_{22} - b_{11} - b_{22}}{b_{11}(b_{22} + 0.22a_{22}) + b_{22}(b_{11} + 0.22a_{11})} =: C_S, \tag{173}$$

in [30] the condition

$$\tau \quad < \quad \frac{D_1 + F_2 - B_1 - E_2}{2(B_1F_2 + D_1E_2 + B_1E_2)} = \frac{a_{11} + a_{22} - b_{11} - b_{22}}{2(b_{11}b_{22} + a_{11}b_{22} + a_{22}b_{11})} =: C_{FR}. \tag{174}$$

Furthermore let us denote the right hand side of the condition in **Theorem 1.10** by C_{GyK}

$$\tau < \frac{a_1 - |b_1|}{|b_0| + 2|c|} = \frac{a_{11} + a_{22} - |b_{11} + b_{22}|}{2b_{11}b_{22} + |a_{11}b_{22} + a_{22}b_{11}|} =: C_{GyK}. \tag{175}$$

Since $a_{11}, a_{22}, b_{11}, b_{22} > 0$, the numerators of C_{FR} and C_{GyK} are equal and

$$2b_{11}b_{22} + |a_{11}b_{22} + a_{22}b_{11}| = 2b_{11}b_{22} + a_{11}b_{22} + a_{22}b_{11} < 2(b_{11}b_{22} + a_{11}b_{22} + a_{22}b_{11}), \tag{176}$$

because

$$a_{11}b_{22} + a_{22}b_{11} < 2(a_{11}b_{22} + a_{22}b_{11}) \tag{177}$$

is true following from the positivity of these constants. This means that $C_{GyK} > C_{FR}$.

Repeatedly, following from the positivity of the constants a_{11}, a_{22}, b_{11} and b_{22} we get that the numerators of C_{GyK} and C_S are equal, but

$$b_{11}(b_{22} + 0.22a_{22}) + b_{22}(b_{11} + 0.22a_{11}) < 2b_{11}b_{22} + a_{11}b_{22} + a_{22}b_{11}, \qquad (178)$$

hence $C_{GyK} < C_S$. On one hand our result in Theorem 1.10 is applicable in general cases, because we have no additional constraints on the sign of the coefficients of the system (123), which means that in this sense our result is better. On the other hand we can increase the upper bound C_S a little bit in the following way. In [31] Stépán used the estimation

$$\sin(x) > -0.22x \qquad (x > 0), \qquad (179)$$

but actually this estimation is not sharp for positive x, cf. **Figure 5**.

If we find a tangent line of the sine function at a certain point x_0, such that this line passes through the origin, i.e. the equation of this line is $y = ax$ with a certain $a < 0$, then we can get a better estimation than (179), namely

$$\sin(x) \geq ax \qquad (x > 0). \qquad (180)$$

We can easily determine the constant $a < 0$ in the following way: the equation of the searched tangent line at x_0 is

$$y = f'(x_0)(x - x_0) + f(x_0) = f'(x_0)x + f(x_0) - f'(x_0)x_0, \qquad (181)$$

where f is the sine function. We would like to find an x_0 such that

$$f(x_0) - f'(x_0)x_0 = \sin(x_0) - \cos(x_0) \cdot x_0 = 0 \qquad (182)$$

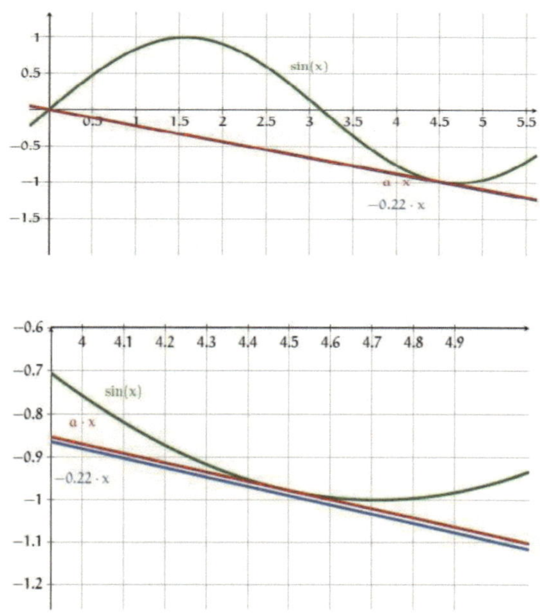

Figure 5.
The estimation of the sine function on the interval $[0, 5.5]$ *and* $[4, 4.9]$.

fulfills, which is true if and only if x_0 is a solution of the equation $\tan(x) = x$. Let x_0 be the solution of this equation, then we find a better linear lower estimation for the function sine on the positive half-line

$$\sin(x) \geq \cos(x_0) \cdot x \qquad (x > 0), \tag{183}$$

i.e. $a := \cos(x_0) \approx -0.21724$.

The proposition of the last theorem of this section is similar to one of Theorem 1.10. In the proof of Theorem 1.10 we used first-order Taylor polynomials to approximate the functions sine and cosine to obtain the estimation (164). But if we use higher-order polynomials, we can get a better result, i.e. a better estimation for τ, such that if τ satisfies the conditions, then the quasi-polynomial (126) is stable.

Theorem 1.11. If the coefficients of the characteristic function (125) fulfill the conditions

$$\left(2 - 6\sqrt{105}\right)a_0 < 59b_0, \qquad -\frac{4}{21}a_1 \leq b_1 < 0, \qquad \frac{32}{3\sqrt{105} - 31}c < b_0 < -32c, \tag{184}$$

and for the delay parameter τ

$$\tau < -\kappa + \sqrt{\kappa^2 - \xi} =: C_2 \tag{185}$$

holds, where

$$\kappa := \frac{B}{2A}, \qquad \xi := \frac{C}{4A} \tag{186}$$

with

$$A := -\left(b_0^2 + 124b_0c + 64c^2\right), \qquad B := 6(a_1(b_0 + 32c) - 4b_1(b_0 + 2c)), \qquad C := 45b_1^2 \tag{187}$$

then the quasi-polynomial (126) is Hurwitz-stable.

Proof: Firstly, let us make the same steps as in the proof of Theorem 1.10 and study the imaginary part of $\Delta(i\omega, \tau)$:

$$\Delta_I(i\omega, \tau) \quad = \quad a_1\omega + b_1\omega\cos(\omega\tau) - b_0\sin(\omega\tau) - 2c\sin(\omega\tau)\cos(\omega\tau). \tag{188}$$

To show that $\Delta_I(i\omega, \tau) > 0$ for each $\omega > 0$ we apply the estimations

$$\sin(x) < x - \frac{x^3}{6} + \frac{x^5}{120}, \qquad \cos(x) > 1 - \frac{x^2}{2} \qquad (x > 0). \tag{189}$$

Then we have

$$\Delta_I(i\omega, \tau) > \omega \cdot P_\tau(\omega), \tag{190}$$

with

$$P_\tau(\omega) := \omega\left(-\omega^4\frac{\tau^5}{120}(b_0 + 32c) + \omega^2\frac{\tau^2}{6}(\tau(b_0 + 8c) - 3b_1) + a_1 + b_1 - \tau(b_0 + 2c)\right). \tag{191}$$

If the conditions (184) are fulfilled, then the coefficient of ω^4 is positive in the polynomial P_τ, moreover if τ satisfies the condition (185) too, then the discriminant of P_τ is negative. Hence with conditions (184) and (185) the inequality $\Delta_I(i\omega, \tau) > 0$ is valid for all $\omega > 0$. Similarly to the previous proof we get by applying the Mikhailov criterion that the quasi-polynomial is asymptotically stable.

In the following example, we are going to show that in some cases the result of Theorem 1.11 is better than the result in [31].

Example 7. Let us consider the following system of delay differential equations

$$\dot{x} = -0.7x - y + 0.01x(\cdot - \tau), \qquad \dot{y} = -0.2x - 0.4y + 0.07y(\cdot - \tau). \tag{192}$$

The characteristic function of (192) is

$$\Delta(z;\tau) := z^2 + 1.1z + \frac{227}{10000} - e^{-z\tau}\left(\frac{2}{25}z + \frac{53}{1000}\right) + e^{2z\tau}\frac{7}{10000} \qquad (z \in \mathbb{C}, \ \tau \geq 0). \tag{193}$$

Let us see what condition gives [31] for the parameter τ. The condition

$$(a_{11} - b_{11})(a_{22} - b_{22}) > a_{12}a_{21} \tag{194}$$

fulfills, hence if

$$\tau < \frac{a_{11} + a_{22} - b_{11} - b_{22}}{b_{11}(b_{22} + 0.22a_{22}) + b_{22}(b_{11} + 0.22a_{11})} \approx 78.1, \tag{195}$$

then the quasi-polynomial (193) is asymptotically stable. Furthermore simple calculations show that following from Theorem 1.11, the system is asymptotically stable for all $\tau \leq 170.07$, which is greater, than (195). In **Figure 6** the phase portrait of the system (192) could be seen with some values of the parameter τ.

3.1 Stability investigation, independently of the delay

We are going to consider the general form of the characteristic function (126):

$$\Delta(z, \tau) = p(z) + q(z)e^{-z\tau} + r(z)e^{-2z\tau} \qquad (z \in \mathbb{C}) \tag{196}$$

where $\deg(r) < \deg(q)$. We assume that if there is not any delay in the system, i.e. $\tau = 0$, then the trivial equilibrium point is asymptotically stable, which is equivalent to the assumption that the polynomial $\Delta(\cdot, 0)$ is Hurwitz stable. We

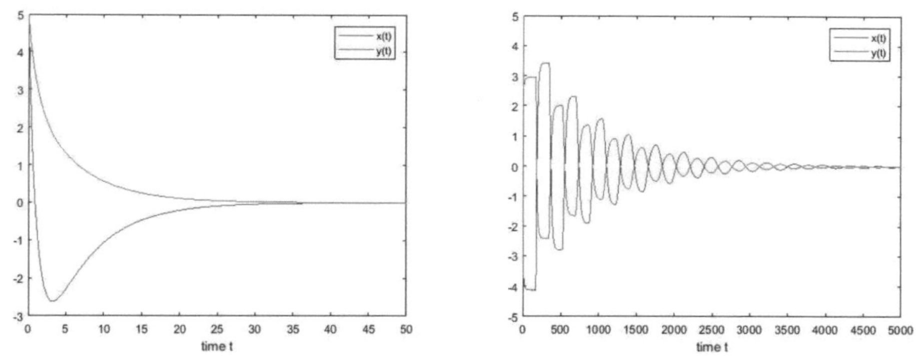

Figure 6.
The solutions of system (192) with $\tau = 0.5$ and $\tau = 170$.

know from [32] that with this assumption the system (and also its trivial equilibrium point) is delay-independently asymptotically stable if and only if for every $\tau > 0$ the quasi-polynomial $\Delta(\cdot, \tau)$ has no non-zero real root on the imaginary axis.

In this chapter we will add a condition to the polynomial p, q and r such that the mentioned property on the root of $\Delta(\cdot, \tau)$ fulfills. In the computations we will follow the idea of [33].

Firstly, let us multiply the equality $\Delta(i\omega, \tau) = 0$ by $e^{i\omega\tau}$, then we can see that the equivalence

$$\Delta(i\omega, \tau) = 0 \iff e^{i\omega\tau}\Delta(i\omega, \tau) = 0 \tag{197}$$

is valid. Let us introduce the notations

$$\begin{aligned} p(i\omega) &=: p_R(i\omega) + ip_I(i\omega), \quad & q(i\omega) &=: q_R(i\omega) + iq_I(i\omega), \\ r(i\omega) &=: r_R(i\omega) + ir_I(i\omega), \quad & \Delta(i\omega, \tau) &=: \Delta_R(i\omega, \tau) + i\Delta_I(i\omega, \tau). \end{aligned} \tag{198}$$

With these notations the characteristic function (196) can be written at $z = i\omega$ in the form

$$\begin{aligned} e^{i\omega\tau}\Delta(i\omega, \tau) = e^{i\omega\tau}p(i\omega) + q(i\omega) + r(i\omega)e^{-i\omega\tau} &= (p_R(i\omega) + ip_I(i\omega))(\cos(\omega\tau) \\ +i\sin(\omega\tau) + (q_R(i\omega) + iq_I(i\omega)) &+ (r_R(i\omega) + ir_I(i\omega))(\cos(\omega\tau) \\ -i\sin(\omega\tau) = \Delta_R(i\omega, \tau) + i\Delta_I(i\omega, \tau). \end{aligned} \tag{199}$$

Let

$$x := \cos\left(\frac{\omega\tau}{2}\right) \text{ and } y := \sin\left(\frac{\omega\tau}{2}\right), \tag{200}$$

then with straightforward calculations we can make the following transformations:

$$\begin{aligned} \Delta_R(i\omega, \tau) &= \cos(\omega\tau)(p_R + r_R) + q_R + \sin(\omega\tau)(r_I - p_I) \\ &= \left(\cos^2\left(\frac{\omega\tau}{2}\right) - \sin^2\left(\frac{\omega\tau}{2}\right)\right)(p_R + r_R) \\ &+\left(\cos^2\left(\frac{\omega\tau}{2}\right) + \sin^2\left(\frac{\omega\tau}{2}\right)\right)q_R + 2\sin\left(\frac{\omega\tau}{2}\right)\cdot\cos\left(\frac{\omega\tau}{2}\right)(r_I - p_I) \\ &=: x(x(q_R + p_R + r_R) + y(r_I - p_I)) + y(x(q_R - p_R - r_R) + y(r_I - p_I)) \\ &=: x(A_x \cdot x + A_y \cdot y) + y(B_x \cdot x + B_y \cdot y) =: Ax + By. \end{aligned} \tag{201}$$

Furthermore we can write the imaginary part in the same way, too:

$$\begin{aligned} \Delta_I(i\omega, \tau) &= \cos(\omega\tau)(p_I + r_I) + q_I + \sin(\omega\tau)(r_R - p_R) \\ &= \left(\cos^2\left(\frac{\omega\tau}{2}\right) - \sin^2\left(\frac{\omega\tau}{2}\right)\right)(p_I + r_I) + \left(\cos^2\left(\frac{\omega\tau}{2}\right) + \sin^2\left(\frac{\omega\tau}{2}\right)\right)q_I \\ &+2\sin\left(\frac{\omega\tau}{2}\right)\cdot\cos\left(\frac{\omega\tau}{2}\right)(r_R - p_R) \\ &= x(x(q_I + p_I + r_I) + y(r_R - p_R)) + y(x(q_I - p_I - r_I) + y(r_R - p_R)) \\ &=: x(C_x \cdot x + C_y \cdot y) + y(D_x \cdot x + D_y \cdot y) =: Cx + Dy. \end{aligned} \tag{202}$$

Hence

$$e^{i\omega\tau}\Delta(i\omega,\tau)=0 \iff \begin{bmatrix} A(\omega\tau) & B(\omega\tau) \\ C(\omega\tau) & D(\omega\tau) \end{bmatrix}\cdot\begin{bmatrix} x \\ y \end{bmatrix}=0. \tag{203}$$

Since the coefficients A, B, C and D in the above matrix are expressed as the linear combination of x and y, we can write expressions $e^{i\omega\tau}\Delta_R(i\omega,\tau)$ and $e^{i\omega\tau}\Delta_I(i\omega,\tau)$ as follows

$$\begin{aligned} e^{i\omega\tau}\Delta_R(i\omega,\tau) &= A_xB_xx^2+(A_xB_y+A_yB_x)xy+A_yB_yy^2 \\ &=: c_0x^2+c_1xy+c_2y^2 \end{aligned} \tag{204}$$

and

$$\begin{aligned} e^{i\omega\tau}\Delta_I(i\omega,\tau) &= C_xD_xx^2+(C_xD_y+C_yD_x)xy+C_yD_yy^2 \\ &=: d_0x^2+d_1xy+d_2y^2. \end{aligned} \tag{205}$$

Then, by dividing the equalities $e^{i\omega\tau}\Delta_R(i\omega,\tau)=0$ and $e^{i\omega\tau}\Delta_I(i\omega,\tau)=0$ by y^2, and introducing a new variable $u:=x/y$, we obtain

$$\Delta_R(i\omega,\tau)=\frac{y^2(c_0u^2+c_1u+c_2)}{e^{i\omega\tau}}, \quad \Delta_I(i\omega,\tau)=\frac{y^2(d_0u^2+d_1u+d_2)}{e^{i\omega\tau}}. \tag{206}$$

Thus, since $e^{i\omega\tau}/y^2\neq 0$, the equation $\Delta(i\omega,\tau)=0$ has no real non-zero root for any given $\tau>0$ if and only of the polynomials f and g have no common real non-zero root, where $f(u)$, resp. $g(u)$ denote the expressions for Δ_R, resp. Δ_I. This is equivalent to that $res(f,g)=det(R[f,g])\neq 0$ or if $res(f,g)=det(R[f,g])=0$, then $\mathrm{discr}[f]<0$ and $\mathrm{discr}[g]<0$, where the resultant of the polynomials f and g is defined as

$$R[f,g]=\det\begin{bmatrix} c_0 & c_1 & c_2 & 0 \\ 0 & c_0 & c_1 & c_2 \\ d_2 & d_1 & d_0 & 0 \\ 0 & d_2 & d_1 & d_0 \end{bmatrix}=(c_0d_0-c_2d_2)^2+(c_2d_1-c_1d_0)(c_0d_1-c_1d_2), \tag{207}$$

and the discriminant of a polynomial $F(u):=au^2+bu+c$ is $\mathrm{discr}[F]:=b^2-4ac$. Hence we have proved the following statement.

Theorem 1.12. The characteristic function (196) has not a non-zero root on the imaginary axis if and only if the polynomial $\Delta(\cdot,0)$ is Hurwitz stable, and $res(f,g)=det(R[f,g])\neq 0$ or if $res(f,g)=det(R[f,g])=0$, then $\mathrm{discr}[f]<0$ and $\mathrm{discr}[g]<0$, where

$$f(u)=c_0u^2+c_1u+c_2 \quad \text{and} \quad g(u)=d_0u^2+d_1u+d_0, \tag{208}$$

where c_0, c_1, c_2, d_0, d_1 and d_2 are defined in (204) and (205).

Example 8. Let us consider again system (166), i.e.

$$\dot{x}=Ax+x(\cdot-\tau)+By-\frac{1}{2}y(\cdot-\tau), \qquad \dot{y}=x+x(\cdot-\tau)+Cy \tag{209}$$

with

$$A := -\left(1+\sqrt{7}\right)/2, \quad B := \left(-4+\sqrt{7}\right)/2, \quad C := \left(-3+\sqrt{7}\right)/2. \tag{210}$$

Straightforward calculations show that for all $\omega \in \mathbb{R}$

$$\mathrm{res}(f,g) = 8\omega^4\left(0.5-\omega^2\right)\left(\omega^2-0.25\right)\cdot 0 = 0, \tag{211}$$

$$\mathrm{discr}(f) = -16\omega^2 < 0, \quad \mathrm{discr}(g) = 16\omega^2\left(\omega^2-0.5\right)^2 \geq 0. \tag{212}$$

Thus, the discriminant of g is not negative, therefore the stability of system (209) changes at some value τ^* of the delay (as we have seen in the previous example).

Figure 7 also shows the changing of the stability of system (209), with $\tau = 0$ the origin is asymptotically stable, but with $\tau = 2$ the origin changes to unstable. The solutions of system (209) with different values of the parameter τ can be seen on **Figure** 7. The stability of system (209) changes if the value of the parameter τ increases.

3.2 Hopf bifurcation

In this subsection we are going to see for which value of the delay τ could change the stability of the system (123). For this purpose we are going to give conditions on the coefficients p, q and r to obtain the value of the delay at which stability switch may occur. Let us assume that for $\omega^* > 0$ the conditions of Theorem 1.12 do not fulfill, i.e. for the polynomials f and g (defined in (206)) the resultant is equal to 0 and the discriminants of f or g is nonnegative. Furthermore let us assume that τ^* is

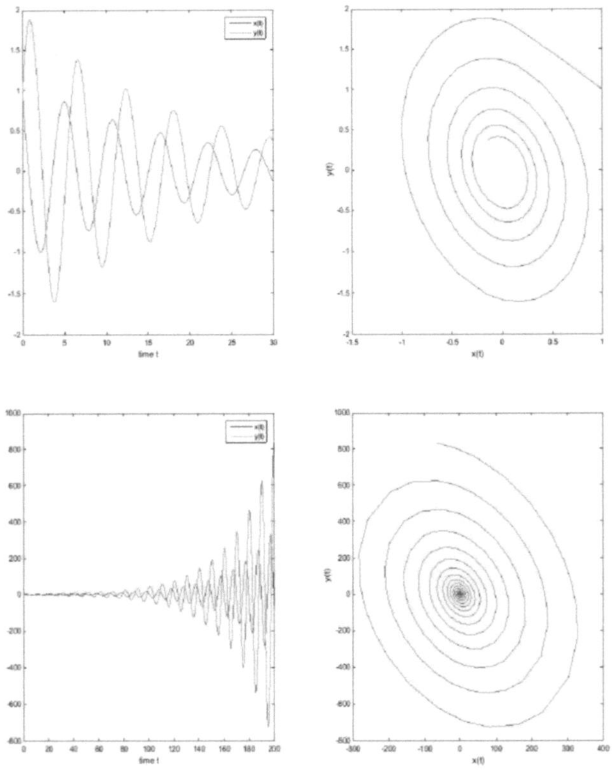

Figure 7.
Example: solutions with $\tau = 0$ *and* $\tau = 2$.

a solution of $\Delta(i\omega^*, \tau) = 0$. Let us denote by $z_3(\tau)$ the root of the quasi-polynomial (196) that assumes $z_3(\tau^*) = i\omega^*$ and the characteristic function $\Delta(\cdot;\tau)$ as a function of the parameter τ by

$$I(z,\tau) := p(z) + q(z)e^{-z\tau} + r(z)e^{-2z\tau} \qquad (z \in \mathbb{C},\ \tau > 0). \tag{213}$$

Thus, we can determine the derivative of z_3 at τ^* by the Implicit Function Theorem (cf. [34]):

$$z_3'(\tau^*) = -\frac{\partial_\tau I(i\omega^*, \tau^*)}{\partial_z I(i\omega^*, \tau^*)} = \frac{\mathcal{E}(z,\tau)}{\mathcal{D}(z,\tau)}\bigg|_{z=i\omega^*, \tau=\tau^*} \tag{214}$$

where $\mathcal{E}(z,\tau) := zq(z) + 2zr(z)e^{-z\tau}$ and $\mathcal{D}(z,\tau) := p'(z)e^{z\tau} + (q'(z) - \tau q(z)) + (r'(z) - 2zr(z))e^{-z\tau}$. To investigate and prove the occurrence of the Hopf bifurcation we have to see the sign of the real part of the above fraction. But since p, q and r are almost arbitrary polynomials, the fraction could be too complicated, that is why we introduce the following notation:

$$\frac{a+ib}{c+id} := \frac{\mathcal{E}(z,\tau)}{\mathcal{D}(z,\tau)}\bigg|_{z=i\omega^*, \tau=\tau^*} \qquad \text{with} \quad \mathfrak{R}\left(\frac{a+ib}{c+id}\right) = \frac{ac+bd}{c^2+d^2}. \tag{215}$$

Since $c^2 + d^2 > 0$, it is enough to consider the sign of $ac + bd$. We are going to use the notations introduced in (198) and along the lines of these we introduce the following notations, too:

$$p'(i\omega) =: p_R'(i\omega) + ip_I'(i\omega), \quad q'(i\omega) =: q_R'(i\omega) + iq_I'(i\omega), \quad r'(i\omega) =: r_R'(i\omega) + ir_I'(i\omega). \tag{216}$$

(For sake of simplicity replacing ω by ω^* we write $P(i\omega^*) =: P$ for $P \in \{p_I, p_R, q_I, q_R, r_I, r_R\}$.)

Computing the exact value of a, b, c and d we have:

$$\begin{aligned}
a &= 2\omega^*\left(r_R \sin(\omega^*\tau^*) - r_I\cos(\omega^*\tau^*) - \frac{1}{2}q_I\right),\\
b &= 2\omega^*\left(r_R \cos(\omega^*\tau^*) + r_I\sin(\omega^*\tau^*) + \frac{1}{2}q_R\right),\\
c &= (p_R' + r_R' - 2\tau^* r_R)\cos(\omega^*\tau^*) + (-p_I' + r_I' - 2\tau^* r_I)\sin(\omega^*\tau^*) + q_R' - \tau^* q_R,\\
d &= (p_I' + r_I' - 2\tau^* r_I)\cos(\omega^*\tau^*) + (p_R' - r_R' + 2\tau^* r_R)\sin(\omega^*\tau^*) + q_I' - \tau^* q_I.
\end{aligned} \tag{217}$$

Thus,

$$\begin{aligned}
ac + bd = 2\omega^*&\Big\{\frac{1}{2}(q_R q_I' - q_I q_R') + \Big[r_R q_I' - r_I q_R' + \frac{1}{2}\big(q_R(p_I' + r_I') - q_I(p_R' + r_R')\big)\Big]\\
&\cdot\cos(\omega^*\tau^*) + \Big[r_I q_I' + r_R q_R' + \frac{1}{2}\big(q_R(p_R' - r_R') + q_I(p_I' - r_I')\big)\Big]\cdot\sin(\omega^*\tau^*)\\
&+ \cos^2(\omega^*\tau^*)\big[r_R(p_I' + r_I') - r_I(p_R' + r_R')\big] + \sin^2(\omega^*\tau^*)\big[r_I(p_R' - r_R') - r_R(p_I' - r_I')\big]\\
&+ \sin(2\omega^*\tau^*)(r_R p_R' + r_I p_I')\Big\}.
\end{aligned} \tag{218}$$

Using the elementary identities

$$\mathfrak{R}(z) \equiv \mathfrak{R}(\bar{z}) \qquad \mathfrak{I}(z) \equiv -\mathfrak{I}(\bar{z}), \qquad \mathfrak{R}(iz) \equiv -\mathfrak{I}(z) \qquad \mathfrak{I}(iz) \equiv \mathfrak{R}(z), \tag{219}$$

furthermore the Euler formula $e^{\pm iz} \equiv \cos(z) \pm i\sin(z)$, we can simplify the enumerator of $\Re(z_3'(\tau))$ as follows $ac + bd = \omega^* \cdot \Im(\mathfrak{A})$, where

$$\mathfrak{A} := \overline{q}q' + 2\overline{r}r' + (2\overline{r}q' + \overline{q}p')e^{i\omega^*\tau^*} + \overline{q}r'e^{-i\omega^*\tau^*} + 2\overline{r}p'e^{2i\omega^*\tau^*}. \tag{220}$$

Therefore, Hopf bifurcation occurs if $\operatorname{sgn}(\Im(\mathfrak{A})) = \pm 1$ holds.

4. Summary

The location of zeros of polynomials and quasi-polynomials as well is crucial in the point of view of the stability of ordinary and retarded differential equations. Namely, if the zeros of the characteristic polynomial of the linearized matrix lie in the open left half of the complex plane, then the constant solution of the particular equation is asymptotically stable. The main task of our work was to depict different methods which allow the investigation of the stability of characteristic (quasi-) polynomials, too. The second objective of this work was in the case of retarded differential equations to treat a method how to estimate the length of the delay for which no stability switching occurs. As an application, we showed a method to detect Hopf bifurcation in ordinary and retarded dynamical systems.

Author details

Sándor Kovács[1]*, Szilvia György[2] and Noémi Gyúró[2]

1 Department of Numerical Analysis, Eötvös Loránd University, Budapest, Hungary

2 Eötvös Loránd University, Budapest, Hungary

*Address all correspondence to: alex@ludens.elte.hu

References

[1] Hale J. Theory of functional differential equations. In: Applied Mathematical Sciences. 2nd ed. Vol. 3. New York-Heidelberg: Springer-Verlag; 1977

[2] Farkas M. Dynamical Models in Biology. New York-Heidelberg: Academic Press; 2001

[3] Gantmacher FR. The theory of matrices. Vol. 1. Translated from the Russian by K. A. Hirsch. Reprint of the 1959 translation

[4] Householder AS. The Theory of Matrices in Numerical Analysis. Reprint of 1964 Edition. New York: Dover Publications, Inc.; 1964

[5] Bissell CC. Stodola, Hurwitz and the genesis of the stability criterion. International Journal of Control. 1989; **50**(6):2313-2332

[6] Tóth J, Szili L, Zachár A. Stability of polynomials. Mathematica in Education and Research. 1998;7(2):5-12

[7] Pontryagin LS. Ordinary differential equations. Translated from the Russian by Leonas Kacinskas and Walter B. Reading, Mass.-Palo Alto, Calif.- London: Counts ADIWES International Series in Mathematics Addison-Wesley Publishing Co., Inc.; 1962

[8] Naulin R, Pabst C. The roots of a polynomial depend continuously on its coefficients. Revista Colombiana de Matemáticas. 1994;**28**(1):35-37

[9] Uherka DJ, Sergott AM. On the continuous dependence of the roots of a polynomial on its coefficients. American Mathematical Monthly. 1977;**84**(5): 368-370

[10] Suter R. Ein kurzer Beweis des Stabilitätskriteriums für kubische Polynome (German) [A short proof of the stability criterion for cubic polynomials]. Zeitschrift für Angewandte Mathematik und Physik. 1983;**34**(6):956-957

[11] Orlando L. Su1 problema di Hurwitz relativo aile parti reali delle mdici di un'equationealgebrica. Mathematische Annalen. 1911;**71**:233-245

[12] Gantmacher FR. The theory of matrices. Vol. 2. Translated from the Russian by K. A. Hirsch. Reprint of the 1959 translation

[13] Jia L. Another elementary proof of the stability criterion of Liénard and Chipart. Chinese Quarterly Journal of Mathematics. 1999;**14**(3):76-79

[14] Kolmanovskiĭ VB, Nosov VR. Stability of functional-differential equations. In: Mathematics in Science and Engineering. Vol. 180. London: Academic Press, Inc.; 1986

[15] Farkas M. Periodic Motions. New York, Inc.: Springer-Verlag; 1994

[16] Younseok C. An elementary proof of the Jury test for real polynomials. Automatica. 2011;**47**(1):249-252

[17] Murray JD. Mathematical biology. I. An introduction. New York: Springer-Verlag; 2002

[18] Kovács S, György S, Gyúró N. On an invasive species model with harvesting. In: Mondaini R, editor. Trends in Biomathematics: Modeling Cells, Flows, Epidemics, and the Environment. Switzerland AG: Springer, Springer Nature; 2020. pp. 299-334

[19] Asada T, Semmler W. Growth and finance: An intertemporal model. Journal of Macroeconomics. 1995;**17**(4): 623-649

[20] Liao H, Zhou T, Tang Y. The chaotic region of Lorenz-type system in the parametric space. Chaos, Solitons and Fractals. 2004;**21**(1):185-192

[21] Lorenz EN. Deterministic nonperiodic flow. Journal of the Atmospheric Sciences. 1963;**20**:130-141

[22] Chen G, Ueta T. Yet another chaotic attractor. International Journal of Bifurcationand Chaos in Applied Sciences and Engineering. 1999;**9**(7): 1465-1466

[23] Lü J, Chen G. A new chaotic attractor coined. International Journal of Bifurcation and Chaos in Applied Sciences and Engineering. 2002;**12**(3): 659-661

[24] Yan Z. Hopf bifurcation in the Lorenz-type chaotic system. Chaos, Solitons and Fractals. 2007;**31**(5): 1135-1142

[25] Bielczyk N, Bodnar M, Foryś U. Delay can be stabilize: Love affairs dynamics. Applied Mathematics and Computation. 2012;**219**:3923-3937

[26] Hals JK, Verduyn Lunel SM. Introduction to Functional-Differential Equations. Berlin, Heidelberg and New York: Springer Verlag; 1993

[27] Churchill RV, Brown JW. Complex Variables and Applications. 8th ed. Boston: McGraw-Hill; 2009

[28] Henrici P. Applied and computational complex analysis. In: Special Functions—Integral Trpansforms—Asymptotics—Continued Fractions. Reprint of the 1977 Original. Vol. 2. New York: Wiley Classics Library. A Wiley-Interscience Publication. John Wiley & Sons, Inc.; 1974

[29] Kearns KD. Stability of nuclear reactor systems having time delays. A dissertation submitted'to the faculty of the Department of Nuclear Engineering in partial fulfillment of the requirements for the degree of Doctor of Philosophy in the Graduate. Tucson, Arizona: College of the University of Arizona; 1971

[30] Freedman HI, Rao SH, V. Stability criteria for a system involving two time delays. SIAM Journal on Applied Mathematics. 1986;**46**:552-560

[31] Stépán G. Retarded dynamical systems: Stability and characteristic functions. Pitman Research Notes in Mathematics Series. 1989;**210**:159

[32] Hale JK, Infante EF, Tsen FSP. Stability in linear delay equations. Journal of Mathematical Analysis and Applications. 1985;**105**(2):533-555

[33] Wu S, Ren G. Delay-independent stability criteria for a class of retarded dynamical systems with two delays. Journal of Sound and Vibration. 2004; **270**:625-638

[34] Simon P. Válogatott fejezetek a matematikából (Hungarian) [Topics in Mathematics]. Budapest: Eötvös Kiadó; 2019

Chapter 2

Some Proposed Problems on Permutation Polynomials over Finite Fields

Mritunjay Kumar Singh and Rajesh P. Singh

Abstract

From the 19th century, the theory of permutation polynomial over finite fields, that are arose in the work of Hermite and Dickson, has drawn general attention. Permutation polynomials over finite fields are an active area of research due to their rising applications in mathematics and engineering. The last three decades has seen rapid progress on the research on permutation polynomials due to their diverse applications in cryptography, coding theory, finite geometry, combinatorics and many more areas of mathematics and engineering. For this reason, the study of permutation polynomials is important nowadays. In this chapter, we propose some new problems in connection to permutation polynomials over finite fields by the help of prime numbers.

Keywords: finite field, permutation polynomial

1. Introduction to permutation polynomials

In this section, we collect some basic facts about permutation polynomials over a finite field that will be frequently used throught the chapter. First it will be convenient to define permutation polynomial over a finite field.

Definition 1. A polynomial $f(x) \in \mathbb{F}_q[x]$ is said to be a permutation polynomial over $\mathbb{F}_q$ for which the associated polynomial function $c \mapsto f(c)$ ia a permutation of $\mathbb{F}_q$, that is, the mapping from $\mathbb{F}_q$ to $\mathbb{F}_q$ defined by f is one–one and onto.

Finite fields are polynomially complete, that is, every mapping from $\mathbb{F}_q$ into $\mathbb{F}_q$ can be represented by a unique polynomial over $\mathbb{F}_q$. Given any arbitrary function $\phi : \mathbb{F}_q \to \mathbb{F}_q$, the unique polynomial $g \in \mathbb{F}_q[x]$ with $\deg(g) < q$ representing ϕ can be found by the formula $g(x) = \sum_{c \in \mathbb{F}_q} \phi(c)\left(1 - (x - c)^{q-1}\right)$, see ([1], Chapter 7).

Two polynomials represent the same function if and only if they are the same by reduction modulo $x^q - x$, according to the following result.

Lemma 1. *[1] For $f, g \in \mathbb{F}_q[x]$ we have $f(\alpha) = g(\alpha)$ for all $\alpha \in \mathbb{F}_q$ if and only if* $f(x) \equiv g(x) \pmod{(x^q - x)}$.

Due to the finiteness of the field, the followings are the equivalent conditions for a polynomial to be a permutation polynomial.

Definition 2. The polynomial $f \in \mathbb{F}_q[x]$ is a permutation polynomial of $\mathbb{F}_q$ if and only if one of the following conditions holds:

i. the function $f : c \mapsto f(c)$ is onto;

ii. the function $f : c \mapsto f(c)$ is one-to-one;

iii. $f(x) = a$ has a solution in $\mathbb{F}_q$ for each $a \in \mathbb{F}_q$;

iv. $f(x) = a$ has a unique solution in $\mathbb{F}_q$ for each $a \in \mathbb{F}_q$.

1.1 Criteria for permutation polynomials

Some well-known criteria for being permutation polynomials are the following.

1.1.1 First criterion for permutation polynomials

The first and in some way most useful, criterion was proved by Hermite for q prime and by Dickson for general q. This criterion has special name what is called Hermite's criterion.

Theorem 3 (Hermite's criterion). *[1] A polynomial $f(x) \in \mathbb{F}_q[x]$ is a permutation polynomial of $\mathbb{F}_q$ if and only if following two conditions hold:*

i. *$f(x)$ has exactly one root in $\mathbb{F}_q$;*

ii. *for each integer t with $1 \le t \le q-2$ and t not divisible by p, the residue $f(x)^t \pmod{(x^q - x)}$ has degree $\le q-2$.*

For the detailed proof, one can see [1]. Above theorem is mainly used to show negative result. The following is a useful corollary for this purpose.

Corollary 4. *There is no permutation polynomial of degree d dividing $q-1$ over $\mathbb{F}_q$.*

Proof. We note that $\deg\left(f^{\frac{q-1}{d}}\right) = q-1$. The proof follows from the last condition of Hermite's criterion.

Remark 5. Hermite's criterion is interesting theoretically but difficult to use in practice.

1.1.2 Second criterion for permutation polynomials

Theorem 6. *[1] Let $f \in \mathbb{F}_q[x]$. Write*

$$D(f) = \left\{ \frac{f(b) - f(a)}{b-a} : a \neq b \in \mathbb{F}_q \right\}.$$

Then $f(x)$ is a permutation polynomial of $\mathbb{F}_q$ if and only if $0 \notin D(f)$.

1.1.3 Third criterion for permutation polynomials

Theorem 7. *[1] The polynomial $f \in \mathbb{F}_q[x]$ is a permutation polynomial of $\mathbb{F}_q$ if and only if*

$$\sum_{c \in \mathbb{F}_q} \chi(f(c)) = 0$$

for all nontrivial additive characters χ of $\mathbb{F}_q$.

1.1.4 Fourth criterion for permutation polynomials

Theorem 8. *[1] Let the trace map* $Tr : \mathbb{F}_{q^n} \to \mathbb{F}_q$ *be defined as* $Tr(x) = x + x^q + \cdots + x^{q^{n-1}}$. *Then the polynomial* $f \in \mathbb{F}_q[x]$ *is a permutation polynomial of* $\mathbb{F}_q$ *if and only if for every nonzero* $\eta \in \mathbb{F}_q$,

$$\sum_{x \in \mathbb{F}_q} \zeta^{\mathrm{Tr}(\eta f(x))} = 0,$$

where $\zeta = e^{\frac{2\pi i}{p}}$ *is a primitive p-th root of unity.*

In what follows, we will discuss some well known classes of permutation polynomials which are commonly used.

1.2 Some well-known classes of permutation polynomials

In this subsection, several basic results on permutation polynomials are presented. Many times, we see that one of these general classes are obtained by simplifying complicated classes of permutation polynomials for proving their permutation nature.

Theorem 9. *[1] Every linear polynomial, that is, polynomial of the form* $ax + b, a \neq 0$ *over finite field is a permutation polynomial.*

Theorem 10. *[1] The monomial* x^n *is a permutation polynomial over* $\mathbb{F}_q$ *if and only if* $\gcd(n, q-1) = 1$.

Theorem 11. *Let* $g(x)$ *and* $h(x)$ *be two polynomials over* $\mathbb{F}_q$. *Then* $f(x) = g(h(x))$ *is a permutation polynomial over* $\mathbb{F}_q$ *if and only if both* $g(x)$ *and* $h(x)$ *permute* $\mathbb{F}_q$.

1.3 Open problems on permutation polynomials

Very little is known concerning which polynomials are permutation polynomials, despite the attention of numerous authors. There are so many open problems and conjectures on permutation polynomials over finite fields but here we are listing few of them.

Open Problem 12. [2] Find new classes of permutation polynomials of $\mathbb{F}_q$.

Although several classes of permutation polynomials have been found in recent years, but, an explicit and unified characterization of permutation polynomials is not known and seems to be elusive today. Therefore, it is both interesting and important to find more explicit classes of permutation polynomials.

Open Problem 13. [2] Find inverse polynomial of known classes of permutation polynomials over $\mathbb{F}_q$.

The construction of permutation polynomials over finite fields is an old and difficult problem that continues to attract interest due to their applications in various area of mathematics. However, the problem of determining the compositional inverse of known classes of permutation polynomial seems to be an even more complicated problem. In fact, there are very few known permutation polynomials whose explicit compositional inverses have been obtained, and the resulting expressions are usually of a complicated nature except for the classes of the permutation linear polynomials, monomials, Dickson polynomials.

Open Problem 14. [2] Find N_d, where $N_d = N_d(q)$ denote the number of permutation polynomials of degree d over $\mathbb{F}_q$.

To date, there is no method for counting the exact number of permutation polynomials of given degree. However, Koyagin and Pappalardi [3, 4], found the asymptotic formula for the number of permutations for which the associated permutation polynomial has degree smaller than $q - 2$.

1.4 Applications of permutation polynomials

The study of permutation polynomials would not complete without mentioning their applications in other area of mathematics and engineering. It is a major subject in the theory and applications of finite fields. The study of permutation polynomials over the finite fields is essentially about relations between the algebraic and combinatoric structures of finite fields. Nontrivial permutation polynomials are usually the results of the intricate and sometimes mysterious interplay between the two structures. Here we mention some applications of permutation polynomials.

1.5 Coding theory

In coding theory, error correcting codes are fundamental to many digital communication and storage systems, to improve the error performance over noisy channels. First proposed in the seminal work of Claude Shannon [5], they are now ubiquitous and included even in consumer electronic systems such as compact disc players and many others. Permutation polynomials have been used to construct error correcting codes. Laigle-Chapuy [6] proposed a conjecture equivalent to a conjecture related to cross-correlation functions in coding theory. In [7], Chunlei and Helleseth derived several classes of p-ary quasi-perfect codes using permutation polynomials over finite fields. In 2005, Carlet, Ding and Yuan [8] obtained Linear codes using planar polynomials over finite fields.

1.6 Cryptography

The advent of public key cryptography in the 1970's has generated innumerable security protocols which find widespread application in securing digital communications, electronic funds transfer, email, internet transactions and the like. In recent years, permutation polynomials over finite fields has been used to design public key cryptosystem. Singh, Saikia and Sarma [9–15] designed efficient multivariate public key cryptosystem using permutation polynomials over finite fields. The same authors used a group of linearized permutation polynomials to design an efficient multivariate public key cryptosystem [16].

Permutation polynomials with low differential uniformity are important candidate functions to design substitution boxes (S-boxes) of block ciphers. S-boxes can be constructed from permutation polynomials over even characteristics [17] with desired cryptographic properties such as low differential uniformity and play important role in iterated block ciphers.

1.7 Finite geometry

Permutation polynomial $f(x) \in \mathbb{F}_q[x]$ is called a complete permutation polynomial if $f(x) + x$ is also a permutation polynomial and an orthomorphism polynomial if $f(x) - x$ is also a permutation polynomial. Orthomorphism polynomials can be used in check digit systems to detect single errors and adjacent transpositions whereas complete permutation polynomials to detect single and twin errors. For more details on complete mappings and orthomorphisms over finite fields, we refer to the reader [3–19]. In addition, complete permutation polynomials are very useful in the study of orthogonal latin squares and orthomorphism polynomials are useful in close connection to hyperovals in finite projective plane. In 1968, planar functions were introduced by Dembowski and Ostrom [20] in context of finite geometry to describe projective planes with specific properties. Since 1991, planar functions have attracted interest also from cryptography as functions with optimal resistance to differential cryptanalysis.

2. Some proposed problems

Let $\mathbb{F}_q$ denotes finite fields with $q = 2^m$ elements. Nowaday permutation polynomials are an interesting subject for study not for only research purposes but also for their various applications in many areas of mathematics and engineering. We refer [21] to the reader for recent advances and contributions to the area.

The rising applications of permutation polynomials in mathematics and engineering from last decade propels us to do new research. Recently, permutation polynomials with few terms over finite fields paying more attention due to their simple algebraic form and some extraordinary properties. We refer to the reader [22–25] for some recent developments. This motivates us to propose some new problems. In this chapter, by the help of prime numbers, we constructed several new polynomials that have no root in μ_{2^m+1} and two of them are generalizations of known ones. The constructed polynomials here may lay a good foundation for finding new classes of permutation polynomials.

Throughout the chapter, for a positive integer d, the set of d-th roots of unity in the algebraic closure $\overline{\mathbb{F}}_q$ of $\mathbb{F}_q$ is denoted by μ_d. That is,

$$\mu_d = \{x \in \overline{\mathbb{F}}_q : x^d = 1\}.$$

For every element $x \in \mathbb{F}_q$, we denote x^{2^m} by $\overline{x}$ in analogous to the usual complex conjugation. Clearly, $x\overline{x}, x + \overline{x} \in \mathbb{F}_q$. Define the unit circle of $\mathbb{F}_q$ as

$$\mu_{2^m+1} = \{x \in \mathbb{F}_q : x^{2^m+1} = x\overline{x} = 1\}.$$

The permutation polynomial of the form $x^r h\left(x^{\frac{q-1}{d}}\right)$ are interesting and have been paid attention, where $h(x) \in \mathbb{F}_q[x]$ with d dividing $q - 1$ and $1 \leq r \leq \frac{q-1}{d}$. The permutation behavior of this type of polynomials are investigated by Park and Lee [26] and Zieve [27].

Lemma 2 ([26, 27]). *Let $r, d > 0$ with d dividing $q - 1$ and $h(x) \in \mathbb{F}_q[x]$. Then $f(x) = x^r h\left(x^{\frac{q-1}{d}}\right)$ permutes $\mathbb{F}_q$ if and only if*

i. $\gcd\left(r, \frac{q-1}{d}\right) = 1$ *and*

ii. $x^r h(x)^{\frac{q-1}{d}}$ *permutes* μ_d.

In view of Lemma 2, the permutation property of $x^r h\left(x^{\frac{q-1}{d}}\right)$ is decided by whether $x^r h(x)^{\frac{q-1}{d}}$ permutes μ_d. In the process to prove that $x^r h(x)^{\frac{q-1}{d}}$ permutes μ_d, first we need to prove that $h(x)$ has no root in μ_d [22]. Thus the polynomials which have no roots in μ_d are interesting and can be used to construct new classes of permutation polynomials. Therefore, it is is both interesting and important to find more polynomials that have no roots in μ_d which play key role in showing the permutation property of $x^r h(x)^{\frac{q-1}{d}}$. For more recent progresses about this type of constructions, we refer [23, 25]. In next section, we also need the following definition.

Definition 15. Two polynomials are said to be conjugate to each other if one is obtained by raising 2^m-th power and multiplying them by the highest degree term of the other.

Next, we propose some new problems by reviewing various recent contributions. The polynomials that have no roots in μ_{2^m+1} play important role in theory of finite fields because these polynomials may give rise to a new class of permutation polynomials.

Let $p \in \{1, 2, \ldots, 2^m - 1\}$, and let the binary representation of p be

$$p = \sum_{k=0}^{m-1} p_k 2^k$$

with $p_k \in \{0, 1\}$. Define the weight of p by

$$w(p) = \sum_{k=0}^{m-1} p_k.$$

We define a polynomial function over $\mathbb{F}_{2^m}$ as

$$L_p(x) = \sum_{k=0}^{m-1} p_k \cdot x^{2^k}.$$

For example,

$$L_{11}(x) = 1 + x + x^3$$

$$L_{13}(x) = 1 + x^2 + x^3$$

$$L_{19}(x) = 1 + x + x^4.$$

We observe that there is a good connection between prime numbers and polynomials that have no roots in μ_{2^m+1} in the sense that most of these polynomials can be derived from prime numbers. In this way, for the prime numbers 11, 13 and 19 we get the polynomials $L_{11}(x), L_{13}(x)$ and $L_{19}(x)$ respectively that have no roots in μ_{2^m+1}. This result is obtained by Gupta and Sharma in [22]. More precisely,

Lemma 3 ([22]). *Let $m > 0$ be integer. Then each of the polynomials $1 + x + x^3, 1 + x^2 + x^3$ and $1 + x + x^4$ have no roots in μ_{2^m+1}.*

Similarly, for the primes 59 and 109, we obtain the same polynomials as in [25] of Xu Guangkui et al.

Lemma 4 ([25]). *Let $m > 0$ be integer. Then each of the polynomials $1 + x + x^3 + x^4 + x^5$ and $1 + x^2 + x^3 + x^5 + x^6$ have no roots in μ_{2^m+1}.*

It is not necessary that all polynomials are obtained from prime numbers. For example, the polynomials $1 + x^3 + x^4$ by Gupta and Sharma in [22] and $1 + x + x^2 + x^4 + x^5$ by Xu Guangkui et al. [25] are obtained corresponding to the number 25 and 55 respectively. In this respect, we propose the following problem.

Problem 16. *Which prime numbers will give polynomials that have no roots in μ_{2^m+1}?.*

The generalization of Lemma 2.2 of [22] corresponding to the polynomials $1 + x + x^3$ and $1 + x^2 + x^3$ are given by the following lemma.

Lemma 5. *For sufficiently large positive integers m and n, each of the polynomials* $1 + x^n + x^{2n-1}$ *and* $1 + x^n + x^{2n+1}$ *have no roots in* μ_{2^m+1}.

Proof. Suppose $\alpha \in \mu_{2^m+1}$ satisfies the equation

$$1 + \alpha^n + \alpha^{2n-1} = 0. \tag{1}$$

Raising both sides of (1) to the 2^m-th power and multiplying by α^{2n-1}, we get

$$1 + \alpha^{n-1} + \alpha^{2n-1} = 0. \tag{2}$$

Adding (1) and (2), we get

$$\alpha^{n-1} + \alpha^n = 0$$

Since $\alpha \neq 0$, which gives $\alpha = 1$. But $\alpha = 1$ does not satisfy (1), a contradiction. Hence $1 + x^n + x^{2n-1}$ has no roots in μ_{2^m+1}. Similarly, we can show that the polynomial $1 + x^n + x^{2n+1}$ has no roots in μ_{2^m+1}.

In particular, we get the following lemma by Gupta and Sharma [22].

Lemma 6 ([22]). *Let* $m > 0$ *be integer. Then each of the polynomials* $1 + x + x^3$ *and* $1 + x^2 + x^3$ *have no roots in* μ_{2^m+1}.

Based on the Lemma 5, we propose the following problem.

Problem 17. *Let* $h_1(x) = 1 + x^n + x^{2n-1}$ *and* $h_2(x) = 1 + x^n + x^{2n+1}$. *Characterize n and r such that the polynomials* $x^r h_1(x)^{2^m-1}$ *and* $x^r h_2(x)^{2^m-1}$ *permutes* μ_{2^m+1}.

By the help of prime numbers below 1000, we obtain the following polynomials that have no roots in μ_{2^m+1}. Most of these polynomials are directly or indirectly associated with prime numbers in the sense that corresponding to either each polynomial or their conjugate polynomial, a prime number can be obtained. The proof of the following lemmas can be done in similar fashion as in [22].

Lemma 7. *For a positive integer m, each of the polynomials* $1 + x + x^2 + x^7 + x^8$, $1 + x + x^6 + x^7 + x^8$, $1 + x + x^3 + x^7 + x^8$, $1 + x + x^5 + x^7 + x^8$, $1 + x + x^4 + x^8 + x^9$, $1 + x + x^5 + x^8 + x^9$, $1 + x^2 + x^3 + x^5 + x^8$, $1 + x^3 + x^5 + x^6 + x^8$, $1 + x + x^3 + x^4 + x^8$, $1 + x^4 + x^5 + x^7 + x^8$, $1 + x^2 + x^3 + x^6 + x^8$, $1 + x^2 + x^5 + x^6 + x^8$, $1 + x^3 + x^4 + x^7 + x^8$, $1 + x + x^4 + x^5 + x^8$, $1 + x^3 + x^4 + x^6 + x^9$, $1 + x^3 + x^5 + x^6 + x^9$, $1 + x + x^2 + x^7 + x^9$, $1 + x^2 + x^7 + x^8 + x^9$, $1 + x^2 + x^4 + x^7 + x^9$, $1 + x^2 + x^5 + x^7 + x^9$ *have no roots in* μ_{2^m+1}.

Lemma 8. *For a positive integer m, each of the polynomials* $1 + x + x^3 + x^5 + x^6 + x^7 + x^8$, $1 + x + x^2 + x^3 + x^5 + x^7 + x^8$, $1 + x + x^2 + x^3 + x^6 + x^7 + x^8$, $1 + x + x^2 + x^5 + x^6 + x^7 + x^8$, $1 + x + x^4 + x^5 + x^6 + x^7 + x^8$, $1 + x + x^2 + x^3 + x^4 + x^7 + x^8$, $1 + x + x^3 + x^5 + x^6 + x^8 + x^9$, $1 + x + x^3 + x^4 + x^6 + x^8 + x^9$, $1 + x + x^3 + x^4 + x^5 + x^8 + x^9$, $1 + x + x^4 + x^5 + x^6 + x^8 + x^9$, $1 + x + x^2 + x^3 + x^7 + x^8 + x^9$, $1 + x + x^2 + x^6 + x^7 + x^8 + x^9$, $1 + x + x^2 + x^4 + x^7 + x^8 + x^9$, $1 + x + x^2 + x^5 + x^7 + x^8 + x^9$, $1 + x^2 + x^3 + x^4 + x^6 + x^7 + x^9$, $1 + x^2 + x^3 + x^5 + x^6 + x^7 + x^9$, $1 + x^2 + x^3 + x^4 + x^5 + x^7 + x^9$, $1 + x^2 + x^4 + x^5 + x^6 + x^7 + x^9$, *have no roots in* μ_{2^m+1}.

Lemma 9. *For a positive integer m, each of the polynomials* $1 + x + x^2 + x^3 + x^4 + x^6 + x^7 + x^8 + x^9$, $1 + x + x^2 + x^3 + x^5 + x^6 + x^7 + x^8 + x^9$ *have no roots in* μ_{2^m+1}.

The above list of polynomials are not complete. However, computational experiments shows that there should be more polynomials. A complete determination of all polynomials with few terms over finite fields seems to be out of reach for the time bing.

Now, we are in condition to propose the following problem in connection to above three lemmas.

Problem 18. *Find new classes of permutation polynomials corresponding to polynomials obtained in Lemmas 7, 8 and 9.*

Classification

AMS 2020 MSC: 11T06.

Author details

Mritunjay Kumar Singh[1]* and Rajesh P. Singh[2]

1 Government Polytechnic, Gaya, Bihar, India

2 Department of Mathematics, Central University of South Bihar, Gaya, Bihar, India

*Address all correspondence to: mmathbhu2012@gmail.com

References

[1] R. Lidl, H. Niederreiter, Finite fields, Vol. 20, Cambridge University Press, 1997.

[2] R. Lidl, G. L. Mullen, When does a polynomial over a finite field permute the elements of the field?, Amer. Math. Monthly 95 (3) (1988) 243–246.

[3] S. Konyagin, F. Pappalardi, Enumerating permutation polynomials over finite fields by degree, Finite Fields Appl. 8 (4) (2002) 548–553.

[4] S. Konyagin, F. Pappalardi, Enumerating permutation polynomials over finite fields by degree ii, Finite Fields Appl. 12 (1) (2006) 26–37.

[5] C. E. Shannon, A mathematical theory of communication, The Bell system technical journal 27 (3) (1948) 379–423.

[6] Y. Laigle-Chapuy, Permutation polynomials and applications to coding theory, Finite Fields and Their Applications 13 (1) (2007) 58–70.

[7] C. Li, T. Helleseth, Quasi-perfect linear codes from planar and apn functions, Cryptography and Communications 8 (2) (2016) 215–227.

[8] C. Carlet, C. Ding, J. Yuan, Linear codes from perfect nonlinear mappings and their secret sharing schemes, IEEE Trans. Inform. Theory 51 (6) (2005) 2089–2102.

[9] W.-S. Chou, Set complete mappings of finite fields, finite fields, coding theory and advances in communications and computing (1991).

[10] W.-S. Chou, H. Niederreiter, Monomials and binomials over finite fields as r-orthomorphisms (2002).

[11] G. L. Mullen, H. Niederreiter, Dickson polynomials over finite fields and complete mappings, Canadian Mathematical Bulletin 30 (1) (1987) 19–27.

[12] H. Niederreiter, K. H. Robinson, Complete mappings of finite fields, Journal of the Australian Mathematical Society 33 (2) (1982) 197–212.

[13] H. Niederreiter, A. Winterhof, Cyclotomic r-orthomorphisms of finite fields, Discrete mathematics 295 (1–3) (2005) 161–171.

[14] Q. Sun, Q. Zhang, A simple proof of a conjecture about complete mapping over finite fields, Journal-Sichuan University Natural Science Edition 35 (1998) 842–847.

[15] Y. Yuan, Y. Tong, H. Zhang, Complete mapping polynomials over finite field $\mathbb{F}_{16}$, in: International Workshop on the Arithmetic of Finite Fields, Springer, 2007, pp. 147–158.

[16] R. P. Singh, B. K. Sarma, A. Saikia, Public key cryptosystem using a group of permutation polynomials, Tatra Mt. Math. Publ. 77 (2020) 1–24.

[17] K. Nyberg, Differentially uniform mappings for cryptography, in: Workshop on the Theory and Application of of Cryptographic Techniques, Springer, 1993, pp. 55–64.

[18] R. P. Singh, A. Saikia, B. Sarma, Little dragon two: An efficient multivariate public key cryptosystem, Int. J. Netw. Secur. Appl. 2 (2) (2010) 1–10.

[19] R. P. Singh, A. Saikia, B. K. Sarma, Poly-dragon: An efficient multivariate public key cryptosystem, J. Math. Cryptol. 4 (4) (2011) 349–364.

[20] P. Dembowski, T. G. Ostrom, Planes of ordern with collineation groups of order n^2, Math. Z. 103 (3) (1968) 239–258.

[21] X. Hou, Permutation polynomials over finite fields—A survey of recent advances, Finite Fields and Their Applications 32 (2015) 82–119.

[22] R. Gupta, R. Sharma, Some new classes of permutation trinomials over finite fields with even characteristic, Finite Fields and Their Applications 41 (2016) 89–96.

[23] Z. Zha, L. Hu, S. Fan, Further results on permutation trinomials over finite fields with even characteristic, Finite Fields and Their Applications 45 (2017) 43–52.

[24] K. Li, L. Qu, X. Chen, New classes of permutation binomials and permutation trinomials over finite fields, Finite Fields and Their Applications 43 (2017) 69–85.

[25] G. Xu, X. Cao, J. Ping, Some permutation pentanomials over finite fields with even characteristic, Finite Fields and Their Applications 49 (2018) 212–226.

[26] Y. H. Park, J. B. Lee, Permutation polynomials and group permutation polynomials, Bulletin of the Australian Mathematical Society 63 (1) (2001) 67–74.

[27] M. E. Zieve, On some permutation polynomials over ${}_q$ of the form $x^r f\left(x^{(q-1)/d}\right)$, Proceedings of the American Mathematical Society 137 (7) (2009) 2209–2216.

Chapter 3

Effectiveness of Basic Sets of Goncarov and Related Polynomials

Jerome A. Adepoju

Abstract

The Chapter presents diverse but related results to the theory of the proper and generalized Goncarov polynomials. Couched in the language of basic sets theory, we present effectiveness properties of these polynomials. The results include those relating to simple sets of polynomials whose zeros lie in the closed unit disk $U = \{z : |z| \leq 1\}$. They settle the conjecture of Nassif on the exact value of the Whittaker constant. Results on the proper and generalized Goncarov polynomials which employ the q-analogue of the binomial coefficients and the generalized Goncarov polynomials belonging to the Dq- derivative operator are also given. Effectiveness results of the generalizations of these sets depend on whether $q < 1$ or $q > 1$. The application of these and related sets to the search for the exact value of the Whittaker constant is mentioned.

Keywords: Basic sets, Simple sets, Effectiveness, Whittaker constant, Goncarov polynomials, Dq operator

1. Introduction

The Chapter is on the effectiveness properties of the Goncarov and related polynomials of a single complex variable. It is essentially a compendium of certain results which seem diverse but related to the theory of the proper and generalized Goncarov polynomials.

Our first set of results deals with simple sets of polynomial [1], whose zeros lie in the closed unit disk U. It is a complement of a theorem of Nassif [1] which resolved his conjecture on the value of the Whittaker constant [2]. We provide also the relation between this problem and the theory of the proper Goncarov polynomials.

Next are results on a generalization of the problem where the polynomials are of the form

$$p_0(z) = 1; \quad p_n(z) = \sum_{k=0}^{n} \begin{bmatrix} n \\ k \end{bmatrix} a_n^{n-k} z^k; \; n \geq 1, \tag{1}$$

and the points $(a_n)_0^\infty$ are given complex numbers with $\begin{bmatrix} n \\ k \end{bmatrix}$ the q-analogue of the binomial coefficient $\binom{n}{k}$. From the results reported, it is shown that the location of the points $(a_k)_0^\infty$ that leads to favorable effectiveness results depends on whether $q < 1$ or $q > 1$. The relation of this problem to the generalized Goncarov polynomials belonging to the Dq-derivative operator is also recorded.

IntechOpen

It is shown that applying the results of Buckholtz and Frank [3] on the generalized Goncarov polynomials $\{Q_n(z;z_0,z_1,\dots,z_{n-1})\}$ belonging to the D_q-derivative operator when $q>1$, leads to the result that, when the points $(z_k)_0^\infty$ lie in the unit disk U, the resulting polynomials fail to be effective.

Consequently, we provide some results on the polynomials $\{Q_n(z;z_0,z_1,\dots,z_{n-1})\}$ when

$$|z_k| \leq q^{-k}; \quad k \geq 0, \tag{2}$$

with the obtained results justifying the restriction (2) on the points $(z_k)_0^\infty$.

Finally, we provide other relevant and related results on the properties of the generalized Goncarov polynomials $\{Q_n(z;z_0,z_1,\dots,z_{n-1})\}$ belonging to the D_q-derivative operator. For a comprehensive and easy reading, background results are provided in the Preliminaries of sections 2.1–2.5.

2. Preliminaries

We record here some background information for easy reading of the contents of the presentation.

2.1 Basic sets and effectiveness

A sequence $\{p_n(z)\}$ of polynomials is said to be basic if any polynomial and, in particular, the polynomials $1, z, z^2, \dots, z^n, \dots$, can be represented uniquely by a finite linear combination of the form.

$$z^n = \sum_{k=0} \pi_{n,k} p_k(z); n \geq 0. \tag{3}$$

The polynomials $\{p_n(z)\}$ are linearly independent.

In the representation (3), let $f(z) = \sum_{n=0}^{\infty} a_n z^n$ be an analytic function about the origin. Substituting (3) into $f(z)$, we have

$$f(z) = \sum_{n=0}^{\infty} a_n z^n = \sum_{n=0}^{\infty} a_n \sum_{k=0} \pi_{n,k} p_k(z).$$

Formally rearranging the terms, we obtain the series

$$\sum_{k=0}^{\infty} p_k(z) \left[\sum_{n=0}^{\infty} a_n \pi_{n,k} \right].$$

We write

$$\prod_k(f) = \sum_{n=0}^{\infty} a_n \pi_{n,k}; k \geq 0.$$

Hence, we obtain the series

$$\sum_{k=0}^{\infty} \prod_k (f) p_k(z),$$

which is called the basic series associated with the function $f(z)$ and the correspondence is written as

$$f(z) \sim \sum_{k=0}^{\infty} \prod_{k} (f) p_k(z). \tag{4}$$

The coefficients $\{\Pi_k(f)\}$ is the basic coefficients of $f(z)$ relative to the basic set $\{p_k(z)\}$ and is a linear functional in the space of functions $\{f(z)\}$.

If $p_n(z)$ is of degree n then the set is called a simple set and is necessarily a basic set.

The basic series (4) is said to represent $f(z)$ in a disk $|z| \leq r$ where $f(z)$ analytic, if the series is converges uniformly to $f(z)$ in $|z| \leq r$ or that the basic set $\{p_n(z)\}$ represents $f(z)$ in $|z| \leq r$.

When the basic set $\{p_n(z)\}$ represents in $|z| \leq r$ every function analytic in $|z| \leq R, R \geq r$, then the basic set is said to be effective in $|z| \leq r$ for the class $\overline{H}(R)$ of functions analytic in $|z| \leq R$.

When $R = r$, the basic set represents, in $|z| \leq r$, every function which is analytic there and we say that the basic set is effective in $|z| \leq r$.

To obtain conditions for effectiveness, we form the Cannon sum

$$w_n(r) = \sum_{k=0} |\pi_{n,k}| M_k(r), \tag{5}$$

where

$$M_k(r) = \max_{|z|=r} |p_k(z)|. \tag{6}$$

From (3), we have that $w_n(r) \geq r^n$,
so that, if we write

$$\lambda(r) = \lim_{n\to\infty} \sup \{w_n(r)\}^{\frac{1}{n}}, \tag{7}$$

$$\lambda(r) \geq r^n. \tag{8}$$

The function $\lambda(r)$ is called the Cannon function of the set $\{p_n(z)\}$ in $|z| \leq r$.

Theorems about the effectiveness of basic sets are due to Cannon and Whittaker (cf. [2, 4, 5]).

A necessary and sufficient condition for a Cannon set $\{p_n(z)\}$ to be effective, in $|z| \leq r$, is

$$\lambda(r) = r. \tag{9}$$

2.2 Mode of increase of basic sets

The mode of increase of a basic set $\{p_n(z)\}$ is determined by the order and type of the set. If $\{p_n(z)\}$ is a Cannon set, its order is defined, Whittaker [2], by

$$w = \lim_{r\to\infty} \lim_{n\to\infty} \sup \frac{\log w_n(r)}{n \log n}. \tag{10}$$

where $w_n(r)$ is given by (5). The type γ is defined, when $0 < w < \infty$, by

$$\gamma = \lim_{r\to\infty} \frac{e}{w}\left[\limsup_{n\to\infty}\{w_n(r)\}^{\frac{1}{n}} n^{-w}\right]^{\frac{1}{w}}. \tag{11}$$

The order and type of a set define the class of entire functions represented by the set.

Theorem 2.2.1 (Cannon [6]).

The necessary and sufficient conditions for the Cannon set of polynomials to be effective for all entire functions of increase less than order p type q is

$$\limsup_{n\to\infty}\left[\left(\frac{epq}{\mathrm{n}}\right)^{\frac{1}{p}}\{w_n(r)\}^{\frac{1}{n}}\right] \leq 1 \text{ for all } r > 0. \tag{12}$$

2.3 Zeros of simple sets of polynomials

The relation between the order of magnitude of the zeros of polynomials belonging to simple sets and the mode of increase of the sets has led to many convergence results, just as that between the order of magnitude of the zeros and the growth of the coefficients has. In the case of the zeros and mode of increase, the approach to achieve effectiveness is to determine the location of the zeros while that between the zeros and the coefficients is to determine appropriate bounds (cf. Boas [7], Nassif [8], Eweida [9]).

2.4 Properties of the Goncarov polynomials

We record in what follows certain properties of the proper and generalized Goncarov polynomials together with the definitions of the q-analogues and the Dq-derivative operator.

The proper Goncarov polynomials $\{G_n(z, z_0, \ldots . z_{n-1})\}$ associated with the sequence $\{z_n\}_0^\infty$ of points in the plane are defined through the relations, Buckholtz ([10], p. 194),

$$G_0(z) = 1,$$

$$\frac{z^n}{n!} = \sum_{k=0}^{n} \frac{z^{n-k}}{(n-k)!} G_k(z, z_0, \ldots, z_{k-1}); \mathrm{n} \geq 1. \tag{13}$$

These polynomials generate any function $f(z)$ analytic at the origin through the Goncarov series

$$f(z) \sim \sum_{k=0}^{\infty} f^k(z_k) G_k(z, z_0, \ldots, z_{k-1}), \tag{14}$$

which represents $f(z)$ in a disk $|z| \leq r$, if it uniformly converges to $f(z)$ in $|z| \leq r$.

In this case, if $f^{(k)}(z_k) = 0, k \geq 0$, the Goncarov series (14) vanishes and $f \equiv 0$.

A consideration of $g(z) = \sin\frac{\pi}{4}(1-z)$, for which $g^{(n)}\{(-1)^n\} = 0$ and $\sum_{n=0}^{\infty} g^{(n)}\{(-1)^n\} G_n(z, 1, -1, ..) = 0$ cf. Nassif [8], shows that the Goncarov series does not always represent the associated function and hence certain restrictions have to be imposed on the points $(z_k)_0^\infty$ and on the growth of the function $f(z)$.

Concerning the case where the points $(z_k)_0^\infty$ lie in the unit disk U, the Whittaker constant W (cf. Whittaker, Buckholtz, [2, 10]), is defined as the supremum of the number c with the following property:

If $f(z)$ is an entire function of exponential type less than c and if each of $f, f', f'', \ldots\ldots$ has a zero in U then $f(z) \equiv 0$.

Buckholtz [10] obtained an exact determination of the constant W. In fact, if we write

$$H_n = \max |G_k(0; z_0, \ldots, z_{n-1}), \tag{15}$$

where the maximum is taken over all sequences $(z_k)_0^{n-1}$ whose terms lie in U, Buckholtz ([10], Lemma 3) proved that $\lim_{n\to\infty} H_n^{1/n}$ exists and is equal to $\sup_{1\le n<\infty} H_n^{1/n}$.

Moreover, if we put

$$\lim_{n\to\infty} H_n^{1/n} = H = \sup_{1\le n<\infty} H_n^{1/n}, \tag{16}$$

Buckholtz ([10], formula 2) further showed that

$$W = \frac{1}{H}. \tag{17}$$

Employing an equivalent definition of the polynomials $\{G_n(z; z_0, \ldots, z_{n-1})\}$ as originally given by Goncarov [11] in the form

$$G_n(z, z_0, \ldots, z_{n-1}) = \int_{z_0}^{z} ds_1 \int_{z_1}^{s_1} ds_2, \ldots, \int_{z_{n-1}}^{s_{n-1}} ds_n; n \ge 1, \tag{18}$$

and differentiating with respect to z, we can obtain

$$G_n^{(k)}(z, z_0, \ldots, z_{n-1}) = G_{n-k}(z, z_k, \ldots, z_{n-1}); \; 1 \le k \le n-1. \tag{19}$$

Writing

$$G_n(0, z_0, \ldots, z_{n-1}) = F_n(z_0, \ldots, z_{n-1}) \; n \ge 1, \tag{20}$$

then (18) yields, among other results,

$$G_n(z; z_0, \ldots, z_{n-1}) = F_n(z; z_0, \ldots, z_{n-1}) - F_n(z; z_1, \ldots, z_{n-1}); \; n \ge 1, \tag{21}$$

and

$$F_n(0, z_1, \ldots, z_{n-1}) = 0; \; n \ge 1. \tag{22}$$

Applying (21) and (22) to (19) we obtain

$$F_k^{(k)}(z_0, z_1, \ldots, z_{n-1}) = -F_{n-k}(z_0, \ldots, z_{n-1}) \tag{23}$$

for $1 \le k \le n-1$, where the differentiation is with respect to the first argument. Expanding $F_n(z_0, \ldots, z_{n-1})$ in powers of z_0, in the form

$$F_n(z_0, \ldots, z_{n-1}) = \sum_{k=0}^{n} \frac{z_0^k}{k!} F_n^{(k)}(0, z_1, \ldots, z_{n-1}),$$

we arrive through (22) and (23) to the formulae of Levinson [12],

$$F_n(z_0, \dots, z_{n-1}) = -\sum_{k=1}^{n} \frac{z_0^k}{k!} F_{n-k}(z_k, \dots, z_{n-1}). \tag{24}$$

Also, differentiating (18) with respect to z_k, we obtain with Macintyre ([13], p. 243),

$$\frac{\partial}{\partial z_k}\left(G_n^{(k)}(z, z_0, \dots, z_{n-1})\right) = -G_k(z, z_0, \dots, z_{k-1})G_{n-k-1}(z_k, z_{k-1}, \dots, z_{n-1}) \tag{25}$$

for $0 \leq k \leq n-1$.

2.5 The q-analogues and D_q derivatives

Let q be a positive number different from 1. The q–analogue of the positive integer n is given by

$$[n] = \frac{q^n - 1}{q - 1}. \tag{26}$$

Also, the q-analogue of $n!$ is

$$[n]! = [n]\,[n-1] \dots [2]\,[1]; n \geq 1; [0]! = 1, \tag{27}$$

and the q-analogue of the binomial coefficient $\binom{n}{k}$ is

$$\begin{bmatrix} n \\ k \end{bmatrix} = \frac{[n]!}{[k]![n-k]!}\ ;\ 0 \leq k \leq n. \tag{28}$$

Moreover, the D_q– derivative operator, corresponding to the number q is defined as follows: *If* $f(z)$ is any function of z, then

$$D_q(\,f(z)) = \frac{f(qz) - f(z)}{z(q-1)}, \tag{29}$$

so that when $f(z) = z^n$, then according to (26), we have $D_q(z^n) = [n]z^{n-1}$ and if $f(z) = \sum_{n=0}^{\infty} a_n z^{n-1}$ is any function analytic at the origin then

$$D_q f(z) = \sum_{n=1}^{\infty} [n] a_n z^{n-1}. \tag{30}$$

In [3] we have a generalization of the Goncarov polynomials as in (13) belonging to the operator D such that for $f(z) = \sum_{n=0}^{\infty} a_n z^n$,

$$D(\,f(z)) = \sum_{n=1}^{\infty} d_n a_n z^{n-1} \tag{31}$$

associated with the sequence $(z_k)_0^{\infty}$, where $e_n = (d_1 d_2 \dots, d_n)^{-1}, e_0 = 1$ and $(d_n)_1^{\infty}$ is a non-decreasing sequence of numbers to obtain

$$\begin{cases} p_0(z) = 1, \\ e_n z^n = \sum_{k=0}^{n} e_{n-k} z_k^{n-k} P_k(z, ; z_0, \ldots, z_{k-1}); n \geq 1. \end{cases} \tag{32}$$

When $d_n = n$, the relations (32) reduce to (6), hence the polynomials $\{p_n(z)\}$ reduce to the proper Goncarov polynomials $\{G_n(z; z_0, \ldots, z_{n-1})\}$. Comparing (30) and (32), Nassif [14] investigated the class of generalized Goncarov polynomials $\{Q_n(z; z_0, \ldots, z_{n-1})\}$ belonging to the Dq- derivative operator when $d_n = [n]$ and $e_n = \frac{1}{[n]!}$ given by,

$$\begin{cases} Q_0(z) = 1 \\ \frac{z^n}{[n]!} = \sum_{k=0}^{n} \frac{z_k^{n-k}}{[n-k]!} Q_k(z\, z_0, \ldots, z_{k-1}); n \geq 1' \end{cases} \tag{33}$$

and the Goncarov series associated with the function $f(z) = \sum_{n=0}^{\infty} a_n z^n$ is

$$f(z) \sim \sum_{k=0}^{\infty} D_q^k f_k(z_k) Q_k(z; z_0, \ldots, z_{k-1}). \tag{34}$$

Writing

$$R_n(z_0, \ldots, z_{n-1}) = Q_n(0; z_0, \ldots, z_{n-1}) \tag{35}$$

so that

$$R_n(0; z_1, \ldots, z_{n-1}) = 0, \ n \geq 1 \tag{36}$$

then we have from, (32) that

$$R_n(z_0, \ldots, z_{n-1}) = -\sum_{k=0}^{n-1} \frac{z^{n-k}}{[n-k]!} R_k(z_0, \ldots, z_{k-1}). \tag{37}$$

Also, Nassif ([14], Lemma 4.1), proved that

$$Q_n(z; z_0, \ldots, z_{n-1}) = R_n(z_0, \ldots, z_{n-1}) - R_n(z; z_1, \ldots, z_{n-1}). \tag{38}$$

We can verify, with Buckholtz ([10], Lemma 1), from the formulae (33), the following:

$$Q_n(\lambda z, \lambda z_0, \ldots, \lambda z_{n-1}) = \lambda^n Q_n(z, z_0, \ldots, z_{n-1}); \ n \geq 1. \tag{39}$$

$$Q_n(z_0, z_0, \ldots, z_{n-1}) = 0; \ n \geq 1. \tag{40}$$

$$D_q Q_n(z, z_0, \ldots, z_{n-1}) = Q_{n-1}(z, z_1, \ldots, z_{n-1}); \ n \geq 1. \tag{41}$$

And hence, by repeated application of D_q, we obtain

$$D_q^k Q_n(z; z_0, \ldots, z_{n-1}) = Q_{n-k}(z; , z_k, \ldots, z_{n-1}); \ 1 \leq k \leq n-1. \tag{42}$$

Expressing $Q_n(z; z_0, \ldots, z_{n-1})$ as a polynomial of degree n in z, then we have from (27), (29) and (42), that

$$Q_n(z; z_0, \dots, z_{n-1}) = \sum_{k=0}^{n} \frac{z^k}{[n]!} R_{n-k}(z_k, \dots, z_{n-1}). \tag{43}$$

The identities (39) and (43) have been obtained, in their general form, in ([3]; formulae (2.5), (2.9)). Also, a combination of (38) and (42) yields

$$D_q^k R_n(0, z_1, \dots, z_{n-1}) = -R_{n-k}(z_k, \dots, z_{n-1}), \tag{44}$$

for $1 \le k \le n-1$, where the differentiation is with respect to the first argument. Expanding $R_n(z_0, z_1, \dots, z_{n-1})$ in powers of z_0, then (36) and (44) imply that

$$R_n(z_0, z_1, \dots, z_{n-1}) = -\sum_{k=1}^{n} \frac{z_0^k}{[k]!} R_{n-k}(z_k, \dots, z_{n-1}). \tag{45}$$

Finally, if we put

$$h_n = \max |R_n(z_0, z_1, \dots, z_{n-1})|, \tag{46}$$

where the maximum is taken over all sequences $(z_k)_0^{n-1}$ and the terms lie in the unit disk U, then Buckholtz and Frank ([3], Corollary 5.2), proved that

$$\lim_{n\to\infty} h_n^{1/n} = h = \sup_{1 \le n < \infty} h_n^{1/n}. \tag{47}$$

Also, in view of the formulae (33), we can verify that, when $q < 1$,

$$h \ge h_2^{1/2} = \left(1 + \frac{[1]}{[2]}\right)^{\frac{1}{2}} > \left(\frac{3}{2}\right)^{\frac{1}{2}} > 1. \tag{48}$$

3. Results on the zeros of simple sets

3.1 Zeros of simple sets of polynomials and the conjecture of Nassif on the Whittaker constant are discussed here

The following result is known for simple sets of polynomials whose zeros all lie in the unit disk.

Theorem A.([1], Theorem 1).

When the zeros of polynomials belonging to a simple set all lying within or on the unit circle the set will be of increase not exceeding order 1 type 1.378.

Using known contributions in the theory of Goncarov polynomials, we show that the alternative form of the above theorem is as follows:

Theorem 3.1.1 ([Nassif and Adepoju [15], Theorem B)

When the zeros of the polynomials belonging to a simple set all lying in the unit disk, the set will be of increase not exceeding order 1 type $\frac{1}{W}$, where W is the Whittaker constant. It is shown also that the result in this theorem is bes t possible.

Indeed, applying the result of Buckholtz ([10], formula 2), the following theorem which resolved the conjecture of Nassif ([8], p.138), is established.

Theorem 3.1.2 ([15], Theorem B)

Given a positive number ε, a simple set $\{p_n(z)\}$ of polynomials, whose zeros all lie in U can be constructed such that the increase of the set is not less than order 1 type H–ε.

For completeness, we give the proof of Theorem 3.1.1 as a revised version of Theorem A.

Proof of Theorem 3.1.1 (Proof of alternative form of Theorem A)

Let $\{b_n\}_1^\infty$ be a sequence of points lying in the unit disk and consider the set $\{q_n(z)\}$ of polynomials given by

$$q_0(z) = 1;\ q_n(z) = (z + b_n)_1^n\ ;\ n \geq 1. \tag{49}$$

Suppose that z^n admits the representation

$$z^n = \sum_{k=0}^{n} \tilde{w}_{n,k} q_{nk}(z). \tag{50}$$

Then multiplying the matrix of coefficients $\left\{ \binom{n}{k} b_n^{n-k} \right\}$ with its inverse $(\tilde{w}_{n,k})$, we obtain

$$-\tilde{w}_{n,0} = \sum_{k=1}^{n} \binom{n}{k} b_n^k \tilde{w}_{n-k,0};\quad n \geq 1.$$

Write.

$$u_n = \frac{\tilde{w}_{n,0}}{n!};\ n \geq 0, \tag{51}$$

then the above relation will give

$$u_n = -\sum_{k=1}^{n} \frac{b_n^k}{k!} u_{n-k};\quad n \geq 1.$$

And to show the dependence of u_n on the points (b_n), this relation can be rewritten as

$$u_n(b_1, b_2, \ldots, b_n) = -\sum_{k=1}^{n} \frac{b_n^k}{k!} u_{n-k}(b_1, b_2, \ldots, b_{n-k}).$$

Comparing this relation with the identify

$$F_n(z_0, z_1, \ldots, z_{n-1}) = -\sum_{k=1}^{n} \frac{z_0^k}{k!} F_n(z_k, \ldots, z_{n-1}),$$

of Levinson [12], we infer that

$$u_n(b_1, b_2, \ldots, b_n) = F_n(b_n, b_{n-1}, \ldots, b_1). \tag{52}$$

Differentiating (50) k times, $k = 1, 2, \ldots, n-1$, we obtain that

$$\tilde{w}_{n,k} = \binom{n}{k} \tilde{w}_{n-k,0}(b_{k+1}, b_{k+2}, \ldots, b_n). \tag{53}$$

Hence, a combination of (15), (16), (20), (51)-(53) leads to the inequality.

$$|\tilde{w}_{n,k}| \leq \frac{n!}{k!} H^{n-k}; \ 0 \leq k \leq n. \tag{54}$$

Observing that $M(q_k; r) \leq (1+r)^k$ for any value of $r \geq 0$, then the Cannon sum of the set $\{q_n(z)\}$ for $|z| = r$ will, in view of (54), be

$$w_n(r) = \sum_{k=0}^{n} |\tilde{w}_{n,k}| \mathrm{M}(q_k; r) \leq n! \, \mathrm{H}^n \exp\left(\frac{1+r}{\mathrm{H}}\right).$$

It follows from (17) that the set $\{q_n(z)\}$ is of increase not exceeding order 1 type $\frac{1}{W}$. The proof is now completed by applying the results of Walsh and Lucas, cf. Marden ([16], pp. 15,46), with (54) and following exactly the same lines of argument as in ([1], pp.109–110), to arrive at the inequality.

$$|\pi_{n,k}| \leq \frac{n!}{k!} \left(H^{n-k}\right). \tag{55}$$

Since $|p_n(z)| \leq (1+r)^n$ in $|z| \leq r$, it follows that the set $\{p_n(z)\}$ is of increase not exceeding order 1 type H = $\frac{1}{W}$.

This completes the proof of the theorem.

3.2 Background and the proof of the conjecture

Before the proof of Theorem 3.2.1, we note that we can take, $\varepsilon < H - 1$.(In fact, according to Macintyre ([13]; p. 241), we have H > $\frac{1}{0.7378}$). Hence it follows from (16) that corresponding to ε, there exists an integer m such that

$$m > (\log \mathrm{H}) / \log\left(1 + \frac{\epsilon}{2\mathrm{H}}\right) \tag{56}$$

such that

$$H_m^{\frac{1}{m}} > H - \frac{\epsilon}{2}. \tag{57}$$

Moreover, from (20), the definition (15) ensures the existence of the points $(a_k)_1^m$ lying in $|z| \leq 1$ such that

$$H_m = |F_m(a_m, a_{m-1}, \ldots, a_1)|. \tag{58}$$

Having fixed the integer m and the sequence $(a_k)_1^m$, the following Lemma is to be first established.

Lemma 3.2.1 ([15], Lemma 3.2).

For any integer $j \geq 1$, write

$$f_j(z_1, z_2, \ldots, z_j) = F(j+1)_{m+j}(a_m, \ldots, a_i; z_j; a_m, \ldots, a_i; z_{j-1}; \ldots, a_m, \ldots, a_i; z_1; a_m, \ldots, a_1) \tag{59}$$

Then, the complex numbers $(\xi_k)_1^\infty$ can be chosen so that

$$|\xi_k| = 1; \quad k \geq 1, \tag{60}$$

and

$$\left| f_j\left(\xi_1, \xi_2, \ldots, \xi_j\right) \right| = H_m^{j+1}; \quad j \geq 1. \tag{61}$$

Proof.
The proof is by induction.
When $j = 1$, we have from (59) that

$$f_1(z_1) = F_{2m+1}(a_m, \ldots, a_1; z_1; a_m, \ldots, a_1).$$

Then the value ξ_1 will be chosen so that

$$|\xi_1| = 1; \left| f_1(\xi_1) \right| = \sup_{|z_1| \leq 1} \left| f_1(z_1) \right|. \tag{62}$$

Applying the identify (25) of Macintyre to $F_{2m+1}(a_m, \ldots, a_1; z_1; a_m, \ldots, a_1)$, we obtain

$$\frac{d}{dz_1}\left(f_1(z_1)\right) = -F_m(a_m, \ldots, a_1)\, G_m(z_1; a_m, \ldots, a_1),$$

so that (20) and (58) imply that

$$\left| f_1'{}'(0) \right| = H_m^2,$$

where the prime denotes differentiation with respect to z_1.
Hence, in view of (62), Cauchy's inequality yields

$$\left| f_1(\xi_1) \right| \geq H_m^2,$$

and the inequality (61) is satisfied for $j = 1$. Suppose then that, for some value $j = k$, the complex numbers $\xi_1, \xi_2, \ldots, \xi_k$ have been chosen satisfying (60) and (61). The numbers ξ_{k+1} will be fixed so that

$$\begin{cases} |\xi_{k+1}| = 1 \\ |F_{k+1}(\xi_1, \xi_2, \ldots, \xi_{k+1})| = \sup\limits_{|z_{k+1}| \leq 1} |F_{k+1}(\xi_1, \xi_2, \ldots, \xi_k, z_{k+1})| \cdot \end{cases} \tag{63}$$

Proceeding in a similar manner as for the Case $j = 1$ and applying the identity (25) of Macintyre with (58), (59) and (61),we can obtain the inequality.

$$\left| f_{k+1}^1(\xi_1, \xi_2, \ldots, \xi_k, 0) \right| \geq H_m^{k+2}, \tag{64}$$

where the prime denotes differentiation with respect to z_{k+1}..
Applying Cauchy's inequality to the polynomial $F_{k+1}(\xi_1, \xi_2, \ldots, \xi_k, z_{k+1})$,we can deduce, using (63) and (64), that

$$|F_{k+1}(\xi_1, \xi_2, \ldots, \xi_k, z_{k+1})| \geq H_m^{k+2}.$$

Hence, by induction, the inequality (61) of the Lemma is established.

We now prove theorem 3.1.2.

The required simple set $\{p_n(z)\}$ of polynomials is constructed as follows:

$$\begin{cases} P_0(z) = 1, \\ p_{j(m+1)}(z) = \left(z + \xi_j\right)^{j(m+1)}; \; j \geq 1, \\ p_{j(m+1)+i}(z) = \left(z + a_j\right)^{j(m+1)+i}; \; 1 \leq i \leq m; j \geq 0, \end{cases} \tag{65}$$

where the points $(a_k)_1^m$ are chosen to satisfy (63) and the numbers $(\xi_k)_1^\infty$ are fixed as in the Lemma.

It follows that the zeros of the polynomials $\{p_n(z)\}$ all lie in the unit disk U.

Also, if z^n admits the unique linear representation.

$$z^n = \sum_{k=0}^{n} \pi_{n,k} p_k(z), \tag{66}$$

and if we write

$$u_n = \frac{\pi_{n,0}}{n!} \; ; \; n \geq 0, \tag{67}$$

then from the relation (52), we deduce from (59) and (65), that

$$u_{(j+1)m+j} = f_j\left(\xi_1, \xi_2, \ldots, \xi_j\right); \; j \geq 1. \tag{68}$$

Now, in view of (66), the Cannon sum of the set $\{p_n(z)\}$ for $|z| = r$, is $w_n(r) > |\pi_{n,0}|$.

Hence, combining (57), (61), (67) and (68) yields

$$w_{(j+1)m+j}(r) \geq \{(j+1)m + j\}! \left(H - \frac{\in}{2}\right)^{(j+1)m}; \; j \geq 1.$$

vIt follows from this inequality and Theorem 3.1 that the order of the set $\{P_n(z)\}$ is exactly 1 and since $H - \frac{\in}{2} > 1$, the type of the set will be

$$\gamma \geq \left(H - \frac{\in}{2}\right)^{\frac{m}{m+1}} \geq \left(H - \frac{\in}{2}\right)^{\frac{m=1}{m}}. \tag{69}$$

In view of the inequality (56), we deduce from (69) that

$$\gamma > H - \in$$

and Theorem 3.1.2 is established.

This settles the conjecture.

4. Generalization

4.1 As a generalization of the above problem, we consider the simple set $\{p_n(z_n)\}$ given by

$$p_0(z) = 1; p_n(z) = p_n(z; a) = \sum_{k=0}^{n} \begin{bmatrix} n \\ k \end{bmatrix} a_n^{n-k} z^k; n \geq 1, \tag{70}$$

where $\begin{pmatrix} n \\ k \end{pmatrix}$ is the q-analogue of the binomial coefficient $\begin{pmatrix} n \\ k \end{pmatrix}$ and $(a_k)_1^\infty$ is a sequence of given complex numbers. The set $\{p_n(z)\}$ is in fact, the q-analogue of the set $\{q_n(z)\}$ in (49). This study is motivated by the fact that this set is related to the generalized Goncarov polynomials belonging to the Dq-derivative operator. Our results show that effectiveness properties of the set.

$\{p_n(z)\}$ depend on whether $q < 1$ or $q > 1$.

We establish the following:

Theorem 4.1.1 ([17], Theorem 1.1)

When the points $(a_k)_1^\infty$ all lie in the unit disk U, the corresponding set $\{p_n(z)\}$ for $q < 1$,will be effective in $|z| \le r$ for $r \ge \frac{h}{1-q}$, where h is as in (47).

Theorem 4.1.2. ([17], Theorem 3.1)

Given $\in > 0$, the points $(a_k)_1^\infty$ lying in $|z| \le 1$ can the chosen so that the correspondence set $\{p_n(z)\}$ of (70) with $q < 1$ will not be effective in $|z| < r$ for $r < \frac{h-\in}{1-q}$..

Theorem 4.1.3 ([17], Theorem 1.2)

When $q > 1$ and

$$|a_k| \le q^{-k} \ ; q \ge 1 \tag{71}$$

the corresponding set $\{p_n(z)\}$ of (70) will be effective in $|z| \le r$ for $r > \frac{q\gamma}{q-1}$, where $\frac{1}{\gamma}$ is the least root of the equation.

$$\sum_{n=0}^{\infty} q^{-n^2} x^n = 2. \tag{72}$$

Theorem 4.1.2 shows that the result in Theorem 4.1.1 is best possible. Also, the restriction (71) on the sequence $(a_k)_1^\infty$ when $q > 1$, is shown to be justified in the sense that if the restriction is not satisfied, the corresponding set $\{p_n(z)\}$ may be of infinite order and not effective.

Proof.

Proof of Theorem 4.1.1 is similar to the first part of Theorem 3.1.1.

Let z^n admits the representation

$$z^n = \sum_{k=0}^{n} \pi_{n,k}(a_1, a_2, \ldots, a_n) p_k(z), \tag{73}$$

then multiplying the matrix of coefficients $\left\{ \begin{bmatrix} n \\ k \end{bmatrix} a^n{}_{n-k} \right\}$ of the set $\{p_n(z)\}$ with the inverse matrix $(\pi_{n,k})$ we obtain

$$\sum_{k=0}^{n} \begin{bmatrix} n \\ k \end{bmatrix} a^n{}_{n-k} \pi_{k,0}(a_1, a_2, \ldots, a_k) = 0; \ n \ge 1.$$

Putting

$$v_0 = 1, v_k = v_k(a_1, a_2, \ldots\ldots, a_k) = \frac{1}{[k]!} \pi_{k,o}(a_1, \ldots, a_{n-k'}), \tag{74}$$

the above relation yields

$$v_n(a_1,\ldots\ldots,a_n) = -\sum_{k=1}^{n}\frac{a^k}{[k]!}v_{n-k}(a_1,\ \ldots,a_{n-k^{\text{`}}}) \tag{75}$$

Comparing the formulae (45) and (75) we infer that

$$v_k(a_1,\ldots\ldots,a_k) = R_k(a_k,\ldots\ldots,a_1). \tag{76}$$

Moreover, operating Dq on the polynomials $\{p_n(z)\}$, we can deduce, from (28) and (29), that

$$D_q(p_k(z;a_k)) = [K]p_{k-1}(z,a_k);\ k\geq 1. \tag{77}$$

Hence, when the operator Dq acts on the representation (73), then (77) leads to the equality

$$\pi_{n,k}(a_1,\ldots.,a_n) = \frac{[n]}{[k]}\pi_{n-1,k-1}(a_1,\ \ldots,a_n),$$

which, on reduction, yields

$$\pi_{n,k}(a_1,\ldots.,a_n) = \begin{bmatrix} n \\ k \end{bmatrix}\pi_{n-k,0}(a_{K+1},\ \ldots,a_n);0\leq k\leq n. \tag{78}$$

Applying (74), (76) and (78), we obtain

$$\pi_{n,k}(a_1,\ldots.,a_n) = \frac{[n]!}{[k]!}R_{n-k}(a_n,\ \ldots,a_{K+1});0\leq k\leq n. \tag{79}$$

Identify (79) is the bridge relation between the set $\{p_n(z)\}$ and the Goncarov polynomials mentioned earlier.

Suppose $q<1$ and assume that

$$r\geq\frac{h}{1-q}. \tag{80}$$

Since $h>1$ as in (47), and restricting the points $(a_k)_1^\infty$ to lie in the unit disk U as in the theorem, it follows from (28) and (80) that

$$\mathrm{M}(p_k;r)\leq(k+1)r^k\ ;k\geq 0. \tag{81}$$

The Cannon sum of the set $\{p_n(z)\}$ for $|z|=r,$ is evaluated from (46), (47), (79), (80) and (81) to obtain

$$w_n(r) = \sum_{k=0}^{n}|\pi_{n,k}|\mathrm{M}(p_k;r)\leq(n+1)^2r^n, \tag{82}$$

from which it follows that the set $\{p_n(z)\}$ is effective in $|z|\leq r$ for $r\geq\frac{h}{1-q}$ and the theorem is established.

5. Proof

5.1 Proof of Theorem 4.1.2

We argue as in the Proof of Theorem 3.1.2. We first obtain an identity similar to (25) of Macintyre using the following Lemma:

Lemma 5.1.1.

For $n \geq 1$ and $k \geq 0$, the following identity holds.

$$\begin{cases} D_{q,z_k} Q_{k+n}(z; z_0, \dots, z_{k+n-1}) \\ = -Q_k(z; z_o, \dots, z_{k-1}) Q_{n-1}(z_k, z_{k+1}, \dots, z_{k+n-1}), \end{cases} \tag{83}$$

where D_{q,z_k} denote the Dq-derivative with respect to z_k.

Proof of Lemma

The proof is by induction.

For $n = 1, k \geq 0$, we have from the construction formulae (33),

$$Q_{K+1}(z; z_0, \dots, z_k) = \frac{z^{k+1}}{[k+1]!}$$
$$- \sum_{j=0}^{k-1} \frac{z_j^{k+1-j}}{[k+1-j]!} Q_j(z; z_0, \dots, z_{j-1}) - z_k Q_k(z; z_0, \dots, z_{j-1}).$$

Hence, operating D_{q,z_k} on this equality, we have that

$$D_{q,z_k} Q_{k+1}(z; z_0, \dots, z_{k-1}) = -Q_k(z; z_0, \dots, z_{k-1}),$$

so that the identity (83) is satisfied for $n = 1, k \geq 0$. Suppose that (83) is satisfied for$n = 1, 2, \dots, m; k \geq 0$. The formulae (33) can be written for $k + m + 1$ in the form,

$$Q_{k+m+1}(z; z_0, \dots, z_{k+m}) = \frac{z^{k+m+1}}{[k+m+1]!} - \sum_{j=0}^{k-1} \frac{z_j^{k+m+1-j}}{[k+m+1-j]!} Q_j\left(z; z_{0,\dots,z_{j-1}}\right)$$
$$- \frac{z^{m+1}_k}{[m+1]!} (Q_k(z; z_0, \dots, z_{k-1})) - \sum_{j=1}^{m} \frac{z_{k+j}^{m+1-j}}{[m+1-j]!} Q_{k+j}(z; z_0, \dots, z_{k+j-1}).$$

Hence, the derivative $D_{q,zk}$ operating on this equation gives, in view of (83),

$$D_{q,z_k} Q_{k+m+1}(z; z_0, \dots, z_{k+m}) = -\frac{z^m}{[m]!} Q_k(z; z_0, \dots, z_{k-1})$$
$$+ \sum_{j=1}^{m} \frac{z_{k+j}^{m+1-j}}{[m+1-j]!} Q_K(z; z_0, \dots, z_{K-1}) \times Q_{j-1}(z_k; z_{k+1}, \dots, z_{K+j-1}).$$

Or equivalently,

$$D_{q,zk} Q_{k+m+1}(z; z_0, \dots, z_{k+m}) = -Q_k(z; z_0, \dots, z_{k-1})$$
$$\times \left[\frac{z_k^m}{[m]!} - \sum_{j=0}^{m-1} \frac{z_{k+j+1}^{m+j}}{[m-j]!} Q_j(z_k; z_{k+1}, \dots, z_{K+j})\right].$$

Hence, formulae (33) imply that

$$D_{q,z_k} Q_{k+m+1}(z; z_0, \dots, z_{k+m}) = -Q_k(z; z_0, \dots, z_{k-1}) Q_m(z_k; z_{k+1}, \dots, z_{K+m}),$$

and the relation (83) is also valid for $n = m + 1$

The Lemma is thus proved by induction. Now, following similar lines paralleling those of the proof of Theorem 3.1.2, we need to establish a Lemma similar to that used for Theorem 3.1.2.

Indeed, observing that $h > 1$ as in (39), the $\in > 0$ of Theorem 4.1.2 can always be picked less than $h - 1$. Also, from (39) it follows that, corresponding to the number $\in$, there exists an integer m for which

$$m > (\log h)/\log\left(1 + \frac{\in}{2h}\right), \tag{84}$$

such that

$$h_m^{\frac{1}{m}} > h - \frac{\in}{2}. \tag{85}$$

Also, from the definition (46) of h_m, the points $(\alpha_i)_1^m$ lying in U can be chosen so that

$$h_m = |R_m(\alpha_m, \dots, \alpha_1)|. \tag{86}$$

With this choice of the integer m and the points $(\alpha_i)_1^m$, the Lemma to be established is the following:

Lemma 5.1.2.

With the notation

$$u_j\left(z_1, z_{2,\dots,} z_j\right) = R_{(j+1)m+j}(\alpha_m, \dots, \alpha_1; z_j; \alpha_m, \dots, \alpha_1; z_{j-1}; \dots; \alpha_m, \dots, \alpha_1; \alpha_m, \dots, \alpha_1), \tag{87}$$

we can choose a sequence $\left(\xi_j\right)_1^m$ of points on $|z| = 1$ such that

$$\left|u_j\left(\xi_1, \xi_2, \dots, \xi_j\right)\right| \geq m^{j+1}; j \geq 1. \tag{88}$$

Proof.

We first observe, from a repeated application of (30), that an analytic function $f(z)$ regular at the origin, can be expanded in a certain disk $|z| \leq 1$ in a series of the form

$$f(z) = \sum_{n=0}^{\infty} \frac{z^n}{[n]!} D_q^n f(0).$$

Hence, by Cauchy's inequality, we have

$$\mathrm{M}(f, r) \geq r\left|D_q f(0)\right|. \tag{89}$$

Applying the usual induction process, we obtain, from (87) for the case $j = 1$, that

$$u_1(z_1) = R_{2m+1}(\alpha_m, \dots \dots, \alpha_1; z_i; \alpha_m, \dots \dots, \alpha_1)$$

Hence the identity (83) yields

$$\begin{aligned} D_q u_1(z_1) &= D_{q,z_1} Q_{2m+1}(0; \alpha_m, \dots \dots, \alpha_1; z_i; \alpha_m, \dots \dots, \alpha_1) \\ &= -R_m(\alpha_m, \dots \dots, \alpha_1) Q_m(z_i, \alpha_m, \dots \dots, \alpha_1). \end{aligned}$$

Therefore, we obtain

$$D_q u_1(0) = R_m^2(\alpha_m, \dots \dots, \alpha_1), \tag{90}$$

where the D_q is operating with respect to z_1.

Pick the number ξ_1, with $|\xi_1| = 1$, such that

$$|u_1(\xi_1)| = \sup\{|u_1(z_1)| : |z_1| = 1\};$$

hence, a combination of (86), (89) and (90) yields

$$|u_1(\xi_1)| \geq h_m^2,$$

and the inequality (88) is satisfied for $j = 1$. The similarity with the proof of Lemma 3.2.1 shows that the proof of this Lemma can be completed in the same manner as that for ealier Lemma.

We can now prove Theorem 5.1.4.

We note that the points $(a_k)_1^\infty$ lying in U which define the required set $\{p_n(z)\}$ of polynomials (70), are chosen as follows:

$$\begin{cases} a_{\,j(m+1)} = \xi_j \\ a_{\,j(m+1)+i} = \alpha_i \;; 1 \leq i \leq m; j \geq 0 \end{cases}, \tag{91}$$

where the points $(\alpha_i)_1^m$ are fixed as in (86) and the sequence $\left(\xi_j\right)_0^\infty$ of points is determined as in Lemma 5.1.2; and the integer m is chosen as in (84) and (85).

If z^n admits the representation (86), then applying (79), (87) and (91) we have that

$$\pi_{(\,j+1)m+j,o} = [(\,j+1)m+i]! u_j\left(\xi_1, \xi_2, \dots, \xi_j\right), j \geq 1, \tag{92}$$

so that, for the Cannon sum of the set $\{p_n(z)\}$ for $|z| = r$, we obtain, from (85), (88) and (92),

$$w_{(\,j+1)m+j}(r) > [(\,j+1)m+j]! \left(h - \frac{\epsilon}{2}\right)^{(\,j+1)m}; \; r > 0.. \tag{93}$$

Since $q < 1$, we have that

$$\lim_{n \to \infty} ([n]!)^{1/n} = \frac{1}{1-q}. \tag{94}$$

Hence, (93) and (94) yield, for the Cannon function,

$$\lambda(r) = \limsup_{n\to\infty}\{w_n(r)\}^{1/n}$$
$$\geq \limsup_{j\to\infty}\{w_{(j+1)m+j}(r)\}^{1/(j+1)m+j}$$
$$\geq \frac{1}{1-q}\left(h-\frac{\in}{2}\right)^{\frac{m}{m+1}};\ r>0.$$

Noting that $h-\frac{\in}{2}>1$, we conclude, from (84), as in the proof of Theorem (50), that

$$\lambda(r)\geq\frac{h-\in}{1-q}\ ;\ r>0,$$

and $\{p_n(z)\}$ will not be effective in $|z|\leq r$ for $r<\frac{h-\in}{1-q}$. This completes the proof.

5.2 Proof of Theorem 4.1.3

Let $\{p_n(z)\}$ be the basic set in (70) with $q>1$. We first justify the statement that if the restriction (71) is not satisfied the corresponding set $\{p_n(z)\}$ may be of infinite order.

For this, we put

$$a_k = t^k\ ;\ k\geq 1, \tag{95}$$

and let t be such that

$$|t| = \beta,\ \frac{1}{q}<\beta<q \tag{96}$$

We claim that, in this case, the corresponding set $\{p_n(z)\}$ will be of infinite order and hence the effectiveness properties of the set will be violated.

Now, in the identity (37), we let

$$z_k = a_{n-k} = t^{n-k};\ 0\leq k\leq n-1,$$

to obtain

$$\sum_{k=0}^{n}\frac{t^{nk}}{[k]!}R_{n-k}\left(t^{n-k},\ldots,t\right)=0; n>0. \tag{97}$$

Put

$$R_j\left(t^j,\ldots,t\right)=t^{\frac{1}{2}j(j-1)}u_j; j\geq 1, \tag{98}$$

so that (97) yields

$$\sum_{k=0}^{n}\frac{t^{\frac{1}{2}k(k+1)}}{[k]!}u_{n-k}=0\ ;\ n>0. \tag{99}$$

Hence, if we put

$$u(z) = \sum_{n=0}^{\infty} u_n z^n, \tag{100}$$

then (97) implies that

$$u(z) = \frac{1}{\varphi(z)}, \tag{101}$$

where

$$\phi(z,t) = \sum_{n=0}^{\infty} \frac{t^{-\frac{1}{2}n(n-1)}}{[n]!} z^n.$$

Since $|t| = \beta < q$, the function $\phi(z,t)$ is entire of zero order and hence it will have zeros in the finite part of the plane.

Let

$$\sigma = \inf\{|z|; \varphi(z) = 0\} < \infty, \tag{102}$$

then from (100) and (101), we have $\limsup_{n\to\infty} |u_n|^{\frac{1}{n}} = \frac{1}{\sigma} > 0$.

Thus, for the Cannon sum of the set $\{p_n(z)\}$, we have, from (79), (96) and (98), that

$$w_n(r) > |\pi_{n,0}| = [n]! \beta^{\frac{1}{2}n(n-1)} |u_n|. \tag{103}$$

Since $q > 1$ and $\beta > \frac{1}{q}$ then, in view of (102), we deduce from (103) that the set $\{p_n(z)\}$ is of infinite order; as claimed.

To prove Theorem 4.1.3 we first note, from (72), that if we put

$$c = \frac{q\gamma}{q-1}, \tag{104}$$

then

$$c > \frac{1}{q-1}. \tag{105}$$

We then multiply the matrix $\left\{ \begin{bmatrix} n \\ k \end{bmatrix} a_n^{n-k} \right\}$ with the inverse$(\pi_{n,k})$ to get

$$\pi_{n,k} = -\sum_{j=k}^{n-1} \begin{bmatrix} n \\ k \end{bmatrix} a_n^{n-j} \pi_{j,k} : \ n > k; \pi_{k,k} = 1. \tag{106}$$

Now, imposing the restriction (71) on the points $(a_k)_1^{\infty}$, we have from (105) and (106) that

$$|\pi_{k+1,k}| \le c.$$

Thus, the inequality

$$|\pi_{m\,k}| \leq c^{m-k} \;;m \geq k, \tag{107}$$

is true for $m = k,\; k+1$.

To prove (107), in general, we observe that, since $q > 1$,

$$\begin{bmatrix} n \\ j \end{bmatrix} \leq q^{\,j(n-j)} \left\{ \frac{q}{q-1} \right\}^{n-j} \;;\; 1 \leq j \leq n. \tag{108}$$

Assume that (107) is satisfied for $m = k, k+1, \ldots, n-1$; then a combination of (71), (72), (104), (106), (107) and (108) leads to the inequality.

$$|\pi_{n,k}| \leq c^{n-k} \sum_{j=1}^{\infty} \left(\frac{q}{c(q-1)} \right)^{j} q^{-j^2} = c^{n-k}.$$

Hence, it follows by induction, that the inequality (107) is true for $m \geq k$.
Noting that

$$\begin{bmatrix} k \\ j \end{bmatrix} = q^{\,j(k-j)} \left\{ \begin{matrix} k \\ j \end{matrix} \right\} \quad q > 1,$$

where $\left\{ \begin{matrix} k \\ j \end{matrix} \right\}$ is the q–analogue of $\begin{pmatrix} k \\ j \end{pmatrix}$, $q^1 = \frac{1}{q} < 1$, we then deduce from (70) and (71), that

$$\begin{aligned} \mathrm{M}(p_k;r) &\leq r^k \sum_{j=0}^{k} \left\{ \begin{matrix} k \\ j \end{matrix} \right\} q^{-j^2} r^{-j} \\ &\leq r^k \sum_{j=0}^{k} \left\{ \begin{matrix} k \\ j \end{matrix} \right\} q^{-\frac{1}{2}j(\,j-1)} (qr)^{-j} \;;\; q > 1. \end{aligned}$$

Appealing to a result of Al-Salam ([18]; formula 2.5), we deduce that

$$\mathrm{M}(p_k;r) \leq r^k \prod_{j=1}^{k} \left(1 + \frac{1}{q^j r} \right) \;;k \geq 1, r > 0. \tag{109}$$

The Cannon sum of the set $\{p_n(z)\}$ for $|z| = r$ can be evaluated from (107) and (109) in the form

$$w_n(r) \leq \left\{ \prod_{j=1}^{n} \left(1 + \frac{1}{q^j r} \right) \right\} \sum_{k=0}^{n} c^{n-k} r^k. \tag{110}$$

Hence, when $r \geq c$ we should have

$$w_n(r) \leq (n+1) \left\{ \prod_{j=1}^{n} \left(1 + \frac{1}{q^j r} \right) \right\} r^n,$$

from which it follows that the set $\{p_n(z)\}$ is effective in $|z| \leq r$ and Theorem 4.1.3 is proved.

6. Other related results

The Goncarov polynomials belonging to the Dq–derivative operator have other properties of interest and worth recording. Hence, we present, in this section, more results regarding the Goncarov polynomials $\{Q_n(z); z_0, \ldots, z_{n-1}\}$ as defined in (84) which belong to the derivative operator Dq and whose points $(z_n)_0^\infty$ lie in the unit disk U for which $q < 1$ or $q > 1$.

When $q < 1$, the result of Buckoltz and Frank ([3]; Theorem 1.2) applied to the derivative operator Dq leads, in the language of basic sets, to the following theorem:

Theorem 6.1 ([19], Theorem 1).

The set of Gancarov polynomials $\{Q_n(z); z_0, \ldots, z_{n-1}\}$ belonging to the Dq operator, with $q < 1$ and associated with the sequence of points $(z_n)_0^\infty$ in U, is effective in $|z| \leq r$ for $r \geq \frac{h}{1-q}$.

Theorem 1.5 of Buckholtz and Frank [3] shows that the result of Theorem 6.1 above is best possible. They also showed that when $q > 1$ the Goncarov polynomials fail to be effective and also, that if $|z| \leq q^{-n}$, no favorable effectiveness results will occur, thus justifying the restriction $|z| \leq q^{-n}$ on the points $(z_n)_0^\infty$.

We also state and prove the following theorem.

Theorem 6.2 ([19], Theorem 2).

Suppose that $q > 1$ and that the points $(z_n)_0^\infty$ satisfy the restriction (111). Then the Goncarov set $\{Q_n(z; z_0, \ldots, z_{n-1})\}$ belonging to the Dq–derivative operator, will be effective in $|z| \leq r$ for $r \geq \frac{hq}{q-1}$ and this result is best possible.

To prove this theorem we put, as in the proof of Theorem (72),

$$q^1 = \frac{1}{q}, \tag{111}$$

so that $q^1 < 1$ and we differentiate between the Goncarov polynomials belonging to the operations Dq and Dq^1 by adopting the notation.

$$Q_n(z; z_0, \ldots, z_{n-1}) \text{ and } p_n(z; z_0, \ldots, z_{n-1}),$$

for these respective polynomials. Thus, the constructive formulae (33) for these polynomials will be

$$Q_n(z; z_0, \ldots, z_{n-1}) = \frac{z^n}{[n]!} - \sum_{k=0}^{n-1} \frac{z_k^{n-k}}{[n-k]!} Q_k(z; z_0, \ldots, z_{k-1}), \tag{112}$$

and

$$P_n(z; z_0, \ldots, z_{n-1}) = \frac{z^n}{\{n\}!} - \sum_{k=0}^{n-1} \frac{z_k^{n-k}}{\{n-k\}!} P_k(z; z_0, \ldots, z_{k-1}), \tag{113}$$

where $[k]!$ and $\{k\}!$ are the respective q and q^1 analogues of the factorial k. With this notation, the following Lemma is to be proved.

Lemma 6.1.

The following identity is true for $n \geq 1$ and $q > 1$:

$$q^{-\frac{1}{2}n(n+1)} Q_n\left(q^n z; q^n z_0, \ldots\ldots, q z_{n-1}\right) = p_n(z; z_0, \ldots\ldots, z_{n-1}). \tag{114}$$

Proof.
We finish note, from the definition of the analogue $[k]!$ and $\{k\}!$, that

$$\frac{q^{n^2}}{\{n\}!} = \frac{q^{\frac{1}{2}n(n+1)}}{\{n\}!};\ n \geq 1, \tag{115}$$

and

$$\frac{q^{(n-k)^2+(n-k)k}}{\{n-k\}!} = \frac{1}{\{n-k\}!} q^{\frac{1}{2}n(n+1)-\frac{1}{2}k(k+1)}\ ;\ 0 \leq k \leq n. \tag{116}$$

Hence, applying the relations (37) and (112) to $\{Q_N(q^n z; q^n z_0, \ldots, q z_n)\}$, we get

$$Q_n(qz, q^n z_0, \ldots, qz_{n-1}) = \frac{q^{n^2}}{\{n\}!} z^n - \sum_{k=0}^{n-1} \frac{q^{(n-k)^2+(n-k)k}}{\{n-k\}!} z_k^{n-1} Q_k(q^k z, q_k z_0, \ldots, q z_{k-1}).$$

Hence, the relations (115) and (116) can be introduced to yield

$$q^{-\frac{1}{2}n(n+1)} Q_n(q^n z, q^n z_0, \ldots, q z_{n-1}) = \frac{z^n}{\{n\}!} - \sum_{k=0}^{n-1} \frac{z_k^{n-k}}{\{n-k\}!} q^{-\frac{1}{2}k(k+1)} Q_k(q^k z, q^k z_0, \ldots, q z_{k-1}). \tag{117}$$

Now, since

$$q^{-1} Q_1(qz; qz_0) = z - z_0 = p_1(z; z_0),$$

the identity (114) is satisfied for $n = 1$.
Moreover, if (114) is valid for $k = 1, 2, \ldots, n-1$, the relations (113) and (117) will give

$$\begin{aligned} q^{-\frac{1}{2}n(n+1)} Q_n(q^n z, q^n z_0, \ldots, q z_{n-1}) &= \frac{z^n}{\{n\}!} - \sum_{k=0}^{n-1} \frac{z_k^{n-k}}{\{n-k\}!} P_k(z; z_0, \ldots, z_{k-1}) \\ &= P_n(z; z_0, \ldots, z_{n-1}), \end{aligned}$$

and hence the Lemma is established.
Proof of Theorem 6.2.
Write

$$z_k = q^{-k} a_k;\ \ k \geq 0, \tag{118}$$

so that the restriction (111) implies that

$$|a_k| \leq 1\ ;\ k \geq 0 \tag{119}$$

Therefore, a combination of (37), (114), (118) yields

$$Q_n(z; z_0, \ldots, z_{k-1}) = q^{-\frac{1}{2}k(k+1)} P_k(z, z_0, \ldots, z_{k-1}). \tag{120}$$

Also, by actual calculation we have that

$$\frac{[n]!}{[n-k]!}q^{-k(n-k)-\frac{1}{2}k(k+1)} = \frac{\{n\}!}{\{n-k\}!} \quad ; \quad 0 \le k \le n \tag{121}$$

Inserting (118), (120) and (121) into (33), we obtain

$$z^n = \sum_{k=0}^{n} \frac{[n]!}{[n-k]!} z_k^{n-k} Q_k(z; z_0, \ldots, z_{k-1})$$

$$= \sum_{k=0}^{n} \frac{\{n\}!}{\{n-k\}!} a_k^{n-k} P_k(z, a_0, \ldots, a_{k-1}),$$

in the sense that each term in the sum on the left hand side of this relation is equal to the corresponding term in the sum on the right hand side.

Hence, if

$$\mathrm{M}_k(r) = \sup_{|z|=r} |Q_k(z; {}_0, \ldots, {}_{k-1})|$$

$$m_k(r) = \sup_{|z|=r} |P_k(z; a_0, \ldots, a_{k-1})|$$

and $\Omega_n(r)$ and $w_n(r)$ are the respective Cannon sums of the sets $\{Q_n(z; z_0, \ldots, z_{n-1})\}$ and $\{P_n(z; a_0, \ldots, a_{n-1})\}$, it follows that

$$\Omega_n(r) = \sum_{k=0}^{n} \frac{[n]!}{[n-k]!} |z_k|^{n-k} \mathrm{M}_k(r) \tag{122}$$

$$= \sum_{k=0}^{n} \frac{\{n\}!}{\{n-k\}!} |a_k|^{n-k} m_k(r) = w_n(r).$$

Since the points $(a_k)_0^\infty$ lie in U, from (119), then applying Theorem 6.1 we deduce from (122) that the set $\{Q_n(z; z_0, \ldots, z_n)\}$ will be effective in $|z| \le r$ for $r \ge \frac{h}{1-q} = \frac{qh}{q-1}$ as to be proved.

To show that the result of the Theorem is best possible we appeal to Theorem 1.5 of Buckholtz and Frank [3] to deduce that the set $\{P_n(z; a_0, \ldots, a_{n-1})\}$ may not be effective in $|z| \le r$ for $r < \frac{qh}{q-1}$.

In view of the relation (122), we may conclude that the set $\{Q_n(z; z_0, \ldots, z_n)\}$ will not be effective in $|z| \le r$ for $r < \frac{qh}{q-1}$ and Theorem 6.2 is fully established.

6.1 The case of Goncarov polynomials with $Z_k = at^k, k \ge 0$

Nassif [14] studied the convergence properties of the class of Goncarov polynomials $\{Q_n(z; z_0, \ldots, z_{n-1})\}$ generated through the qth derivative described in (33) where now, $z_k = at^k, k \ge 0$ and a and t are any complex numbers. By considering possible variations of t and q, it was shown that except for the cases $|t| \ge 1, q < 1$ and $|t| > \frac{1}{q}; q > 1$, all other cases lead to the effectiveness of the set $Q_n(z; a, at, \ldots at^{n-1})$ in finite circles ([14]; Theorems 1.1, 1.2, 1.3, 3.2, 3.3).

6.2 Quasipower basis (QP-basis)

Kazmin [20] announced results on some systems of polynomials that form a quasipower basis, (QP-basis), in specified spaces. These include the systems of Goncarov polynomials and of polynomials of the form:

$$\{(z+\alpha_n)^n\}, n = 0, i, 2 \ldots ; \alpha_n \in [-1, 1]. \tag{123}$$

For full details of QP-basis and some of the results announced, cf. ([20]; Corollaries 3, 4).

Of interest is his results that the system in (123), for arbitrary sequence $\{a\}_0^\infty$ of complex numbers with $|a_n| \le 1$, forms a QP- basis in the space $[1, \sigma]$, for $0 < \sigma < W$ and in the space $[1, \sigma)$, for $0 < \sigma \le W$, where W = 0.7377 is the Whittaker constant. This value of W = 0.7377 is attributed to Varga [21]. He also added that Corollaries 3 and 4 contain known results in [5, 9, 15, 22, 23].

7. Conclusions

The chapter presents a compendium of diverse but related results on the convergence properties of the Goncarov and Related polynomials of a single complex variable. Most of the results of the author (or joint), have appeared in print but are here presented in considerable details in the proofs and in their development, for easy reading and assimilation. The results of other authors are summarized with related and relevant ones mentioned to complement the thesis of the chapter. Some recent works related to the Goncarov and related polynomials, cf. [24–29], which provides further applications are included in the references.

The comprehensiveness of the presentation is for the needs of those who may be interested in the subject of the Goncarov polynomials in general and also in their application to the problem of the determination of the exact value of the Whittaker constant, a problem that is still topical and challenging.

Acknowledgements

I acknowledge the mentorship of Professor M. Nassif, (1916-1986), who taught me all I know about Basic Sets. I thank Dr. A. A. Mogbademu and his team for typesetting the manuscript at short notice and also the Reviewer for helpful comments which greatly improved the presentation.

No conflict of interest

The author declares no conflict of interest.

Author details

Jerome A. Adepoju
Formerly of the Department of Mathematics, University of Lagos, Akoka-Yaba, Lagos, Nigeria

*Address all correspondence to: jadi1011@yahoo.com

References

[1] Nassif M. Zeros of simple sets of polynomials. Duke Math. J. 1952; 19 (1): 107-113.

[2] Whittaker JM. Interpolatory Function Theory. Cambridge; 1945

[3] Buckholtz JD, Frank JL. Whittaker Constants. Proc. Lond. Math. Soc. 1971; 3(23): 348-371.

[4] Cannon TW. On the convergence of series of polynomials. Proc. Lond. Math. Soc. 1937; 2(43): 248-364.

[5] Whittaker JM. Sur les series des polynomes quelcuonques, Gauthier-Villars, Paris. 1949.

[6] Cannon TW. On the convergence of integral functions by general basic series. Math. Zeit. 1949; 45:185-205

[7] Boas Jr. R P. Basic sets of polynomials I. Duke Math. J. 1948; 15: 717-724.

[8] Nassif M. Products and Zeros of Basic sets. Univ. Assiut Monograph Series. 1958; 50 (2); p.138.

[9] Nassif M. On the mode of increase of simple sets of polynomials of given zeros. Proc. Math. Phys. Soc. Egypt. 1952; 4: 29-36.

[10] Buckholtz JD. The Whittaker constant and successive derivatives of entire functions. J. Approx. Theory, 1970; 3: 194-212.

[11] Goncarov W. Recherches Sur. Les derives successive de fonctions analytiques' Anall. Scient. de L' EcoleNormale Superieures. 1930; 47:1-78.

[12] Levinson N. The Gontcharoff polynomials. Duke Math. J. 1944; 11: 729-733.

[13] Macintyre S S. On the Zeros of successive derivatives of integral functions. Trans. Amer. Math. Soc. 1949; 67: 241-251.

[14] Nassif M. On a class of Generalized Goncarov polynomials. J. Approx. Theory. 18: 329- 345.

[15] Nassif M, Adepoju JA. Zeros of polynomials belonging to simple sets. Natn. Acad. Sci. Lett. 1979; 1(16): 223-224.

[16] Marden M. The Geometry of the Zeros of a polynomial in a complex variable. Amer. Math. Soc. Math. Surv. 3; 1958 :NewYork.

[17] Nassif M, Adepoju J A. A polynomial set Related to Goncarov polynomials. Indian J. Pure Appl. Math 1980; 1 (12): 1665-1672.

[18] AL- Salam, W. A., q- analogue of Cauchy formula. Proc. Amer. Math. Soc. 1966; 17: 616-621.

[19] Adepoju JA. On the convergence of a class of Generalized Goncarov polynomials. Soochow J. Math. 1995; 21 (1): 71-80.

[20] Kaz'min Yu A. Perturbed Appell polynomials and systems of functions Associated with them. Soviet. Math. Dokl. 1985; 31(3): 452-455.

[21] Varga RS. Topics in polynomials and rational interpolation and Approximation. Sem. Math. Sup. 1982. Presses Univ. Montreal, Montreal.

[22] Eweida, M; A note on Abel's polynomials. Proc. Math. Phys. Soc. Egypt. 1959; 5 (3): 63-66

[23] Kaz'min Yu. Vestnic Moscow Univ. Ser. I. Math. Mekh. 1978; 2: 88-90, English Transl. in Moscow Univ. Math. Bull. 1978; 33: 1978.

[24] Eweida MA. Proc. Math. Phys. Soc. Egypt. 1959; 22: 83-88.

[25] Adeniran A, Yan C. Goncarov polynomials in partition lattice and and exponentials families. Advanced in Appl. Math. Elsevier. May 2021.

[26] Lorentz R, et al. Generalized Goncarov polynomials. In Butler J. Cooper and G. Hierlbert (eds). Cambridge Univ. P. Press . 2018.

[27] Wu R. Abel- Goncarov Type multiquadric Quasi- interpolation operators with higher approximation order. Hindawi J. Math. July 2021.

[28] Anjorin A. Akinbode S. Chebychev polynomials of the first kind and the Whittaker Constant. Global J. Sci. Frontier Research: For Math. and Decision Sci. 2015 (2):Version 10

[29] Yacubovich S. On some properties of Abel- Goncerov polynomials and the Casas- Alvero problem. Integral Transforms and special functions. 2016; 27 (8).

Chapter 4

Irreducible Polynomials: Non-Binary Fields

Mani Shankar Prasad and Shivani Verma

Abstract

Irreducible polynomials play an important role in design of Forward Error Correction (FEC) codes for data transmission with integrity and automatic correction of data, as for example, *Low-Density Parity Check codes*. The usage of irreducible polynomials enables construction of non-prime-order finite fields. Most of the irreducible polynomials belong to binary Galois field. The important analytical concept is optimisation of irreducible polynomials for use in FECs in nonbinary Galois (NBG) field, leading to the development of an algorithm for LDPC that can work with nonbinary Galois fields. According to studies, the Tanner graph for 'nonbinary Low Density Parity Check' codes might get sparser as the field's dimensions rise, ensuring that they do much better than their binary counterparts. A detailed discussion of representation of nonbinary irreducible polynomial and the computations involved have been illustrated. The concept has been tried for NB-LDPC codes. To prove the notion, computational complexity is found from different parameters such as performance of error correction capability, complexity cost and simulation time taken. Such detail study makes the NBG fields-based FEC very suitable for high-speed data transmission with self-error correction.

Keywords: irreducible polynomial, nonbinary Galois field, FEC, LDPC, Galois field

1. Introduction

Polynomials are algebraic expressions that consist of variables and coefficients. Indeterminates are another word for variables. Polynomial expressions can undergo a variety of mathematical operations, but they cannot be split by the variable.

The Greek phrases 'poly' and 'nomial', which imply 'many' and 'period', respectively, from the word Polynomial. A polynomial is a mathematical equation that is formed by multiplying the sum of terms in one or more variables by coefficients.

For example

$$5x^3 + \frac{7}{4}x^2 - \frac{2}{3}x + 1 \quad (1)$$

is a polynomial in single variable.

$$\sqrt{2}x^3y^2 + 6x^2y + 5xy + \sqrt{3} \quad (2)$$

is a polynomial in two variables.

For the polynomial, $a_n x^n + a_{n-1}x^{n-1} + \ldots \ldots .. a_0$, the degree is defined to be n if $a_n \neq 0$. The degree of the polynomial $\sum a_{ij}x^i y^j$ in two variables x, and y is given by $\max\{i + j \mid a_{ij} \neq 0\}$.

Given two polynomials $f(x)$ and $g(x)$, there exist unique quotient and remainder polynomials $q(x)$ and $r(x)$, such that

$$f(x) = q(x)g(x) + r(x), \textit{where degree of } r(x) < g(x) \tag{3}$$

If $gcd[f(x), g(x)] = d$, then there exists two polynomials $p(x), q(x)$ such as

$$d = p(x)f(x) + q(x)g(x) \tag{4}$$

1.1 Irreducible polynomials

Irreducible polynomials are considered as the basic constituents of all polynomials.

A polynomial of degree n ≥ 1 with coefficients in a field F is defined as irreducible over F in case it cannot be expressed as a product of two non-constant polynomials over F of degree less than n.

Example 1:

Consider the x^2–2 polynomial. There are no zeroes in $x^2 - 2$ over Q. This is the same as asserting that $\sqrt{2}$ is not rational [1].

In case$x^2 - 2$ is reducible, we may write x^2–2 = g(x)h(x), where g(x) and h(x) are both fewer than two degrees. Because the LHS has a degree of two, the sole option is that both g(x) and h(x) have a degree of one. x^2–2 has a zero in Q in this example, which is a contradiction. As a result, x^2–2 is irreducible over Q. x^2–2 = (x – 2) (x + 2), and on the contrary, is reducible over R, $x^2 - 2 = \left(x - \sqrt{2}\right)\left(x + \sqrt{2}\right)$.

Example 2: $x^4 + x + 1$ is an irreducible polynomial having degree 4 over *GF* (2) but $x^4 + x^3 + x^2 + 1$ is not irreducible because $x^4 + x^3 + x^2 + 1 = (x^3 + x + 1)(x + 1)$.

Every polynomial of degree one is irreducible. The polynomial $x^2 + 1$ is irreducible over R but reducible over C.

Gauss's lemma.

A polynomial $f \in Z[x] \subseteq Q[x]$ of the form

$$f(x) = x^n + a_{n-1}x + n - 1 + \ldots + a_1 x + a_0$$

is irreducible in Q[x] iff it is irreducible in Z[x]. More precisely, if f(x) ∈ Z[x], then f(x) can be factorised and represented as multiplication of two polynomials of lesser degrees r and s in Q[x] iff it has such a factorisation with polynomials of similar degrees r and s in Z[x].

Eisenstein irreducibility criterion (1850).

Suppose that $f(x)$is the polynomial with coefficients in the ring Z of integers, given by $f(x) = c_n x^n + \cdots + c_1 x + c_0$and p is a prime which satisfies

1.p does not divide c_n;

2.p divides $c_{n-1}, \ldots, p_1, p_0$;

3.p^2 does not divide c_0;

Then, $f(x)$ is irreducible over field Q of rational numbers. [2]

Dumas criterion (1906).

Let F(x) $= a_n x^n + a_{n-1}x^{n-1} + \ldots \ldots ..a_0$ be a polynomial with coefficients in Z [2]. Suppose there exists a prime p whose exact power p^{r_i} dividing a_i (where $r_i = \infty$ if $a_i = 0$), $0 \leq$ i $\leq$ n, satisfy.

- $r_n = 0$,
- $(r_i/n - i) > (r_0/n)\, for\; 1 \leq i \leq n-1$ and
- $gcd(r_0, n)$ equals 1.

Then, F(x) is irreducible over Q.

Note that Eisenstein's criterion is a special case of Dumas criterion with $r_0 = 1$.

1.2 Monic polynomials

A polynomial $x^n + a_{n-1}x^{n-1} + \ldots \ldots . + a_1 x + a_0$ in which the coefficient of the highest-order term is 1 is called a monic polynomial. For example, $x^2 + 3$ is monic but, $7x^2 + 3$ is not monic (the highest power of x^2 has a coefficient of 7, not 1).

Let F_q stand for the finite field of q elements, where q is the prime number. Gauss' formula, in general, gives the number of monic irreducible polynomials $M_n(p)$ of degree n over the finite field F_q. [3]

$$M_n(p) = \frac{1}{n}\sum_{d/n} \mu(d/n) q^d \tag{5}$$

where $\mu(r)$ denotes the Mobius function and d includes all positive divisors of n. d also includes 1 and n. Also, $\mu\ (1) = 1$. The value of $\mu(d)$ calculated at a product of distinct primes is 1 if number of factors is even and it is equal to -1 if the number of factors is odd. For all other natural numbers $\mu(d) = 0$, that is,

$$\mu(d) = \begin{cases} 1, & d \text{ is squarefree positive integer with an even number of prime factors} \\ -1, & d \text{ is square freepositive integer with an odd number of prime factors} \\ 0, & d \text{ has a squared prime factor} \end{cases}$$

In particular, $\mu(d) = -1$ for all primes p.

Example: $GF(256) = GF(2^8)$. That is, $p = 2\ and\ n = 8$.

$$Mn(p) = \frac{1}{8}\sum_{d/8} \mu(8/n) 2^d = \frac{1}{8} \sum_{d\varepsilon(1,2,4,8)} \mu\left(\frac{8}{n}\right) 2^d$$

1.3 Primitive polynomials

A 'primitive polynomial' has its roots as primitive elements in the field $GF(p^n)$. It is an irreducible polynomial of degree d. It can be proved that there are $\varnothing\left(p^d - 1/_d\right)$ number of primitive polynomials, where $\varnothing$ is Euler phi-function. For example, if p = 2, d = 4, $\varnothing\left(2^4 - 1/_4\right)$ is 2, so there exist exactly two primitive polynomials of degree 4 over GF (2). The MATLAB function pol = gfprimfd(m,opt,p) searches for one or more primitive polynomials for $GF(p^n)$, where p represents a number that is prime and n is greater than 0. The MATLAB code below seeks primitive polynomials for GF (81) having various other properties.

p = 3; n = 4;
pol1 = gfprimfd(m,'min',p)
pol2 = gfprimfd(m,3,p)
pol3 = gfprimfd(m,4,p)
The output is as shown below:
pol1 =
2 1 0 0 1
pol2 =
2 1 0 0 1
2 2 0 0 1
2 0 0 1 1
2 0 0 2 1
pol3 = [ ]

Because no primitive polynomial for $GF(81)$ has exactly four nonzero terms, pol3 is empty. Also, pol1 represents a single three-term polynomial, whereas pol2 represents all of the three-term primitive polynomials for GF (81).

Further, *Primpoly(m)* gives the primitive polynomial for $GF(2^m)$, where m is an integer between 2 and 16. The output is an integer, whose binary representation represents the polynomial coefficients.

For $GF(2^m)$, *primpoly(m,opt)* returns one or more primitive polynomials. As seen in **Table 1**, the output depends on the argument opt. The output argument (an integer represented in binary format) represents the coefficients of the relevant polynomial. There is no element in the output if no primitive polynomial satisfies the conditions [4].

m = 4;
defaultprimpoly = primpoly(m)
allprimpolys = primpoly(m,'all')
i1 = isprimitive(25)
i2 = isprimitive(21)
The output is as shown below:
Primitive polynomial(s) =
D^4+D^1+1
defaultprimpoly =
19
Primitive polynomial(s) =
D^4+D^1+1
D^4+D^3+1
allprimpolys =
19
25
i1 = logical

opt	Meaning of pr
'min'	One primitive polynomial for GF(2^m) having the smallest possible number of nonzero terms
'max'	One primitive polynomial for GF(2^m) having the greatest possible number of nonzero terms
'all'	All primitive polynomials for GF(2^m)
Positive integer k	All primitive polynomials for GF(2^m) that have k nonzero terms

Table 1.
Different options for argument 'opt' in primpoly (m,opt).

=1
i2 = logical
=0

The MATLAB function *isprimitive(a)* gives the output as 1 if a represents a primitive polynomial for the Galois field $GF(2^m)$, and 0 otherwise [4].

2. Role of primitive polynomials in nonbinary LDPC codes [12]

Error control codes find applications in the transmission and storage of vast amounts of error-prone data. Mostly, binary and nonbinary channels use ECC codes such as BCH codes, Reed Solomon codes [5], and Low-Density Parity Check codes. Berlekamp and Massey, after Peterson, created powerful algorithms that proved possible with the use of latest digital techniques. In addition to this, the usage of primitive polynomials and the Galois field offered these routines a structured and systematic approach.

LDPC codes got an initial explanation from Gallager in 1963. A parity check matrix (PCM) having sparse characteristics with a minimal number of nonzero components feature LDPC codes. These codes were overlooked until the mid-1990s despite their superior performance due to their decoding complexity, which exceeded the capacity of then electronic systems. When Mackay et al. reviewed the LDPC codes in 1996 [6], they discovered that when decoded using probabilistic soft choice decoding methods, they obtain performance close to the Shannon limit. Gallager later proposed nonbinary LDPC (NB-LDPC) codes by extending the concept of LDPC codes to nonbinary alphabets.

The NB-LDPC codes are defined for Galois fields of order, strictly higher than 2. They are considered as a good alternative to LDPC codes (Refer **Figure 1**) because:

- The graph for an NBLDPC code is frequently less dense than the graph for a binary-LDPC code for the same code rate and binary length [7]. As a result, the topological features of the NB-LDPC code plot have improved, with a larger girth and fewer halting and trapping sets.

- For a binary Low-Density Parity Check code sent with a high-order modulation, the Maximum a-posteriori (MAP) de-mapper produces probability at the binary level for the decoder. This means that decoder input messages are associated even when the cycles are absent. If the LDPC code is defined in the same or higher order of Galois field as the order of modulation, the NB-LDPC decoder is initialised with non-correlated messages. NB-LDPC codes outperform other high-order modulation codes as a consequence.

- Better AWGN channel performance: Decoding techniques in [8] push NB-LDPC codes close to capacity limits, especially at high rates for the AWGN channel. Li *et al.* [9] looked at the result of implementing NB-LDPC codes over AWGN channel. The decoding capabilities of NB-LDPC codes demonstrated that the Shannon capacity of the binary AWGN channel may be achieved with suitably long code words by simply increasing the field order q of the NB-LDPC code (**Figure 1**).

2.1 Galois fields

A Galois field is a finite field with a finite order, which is either a prime number or the power of a prime number. A field of order $n^p = q$ is represented as $GF(n^p)$. A

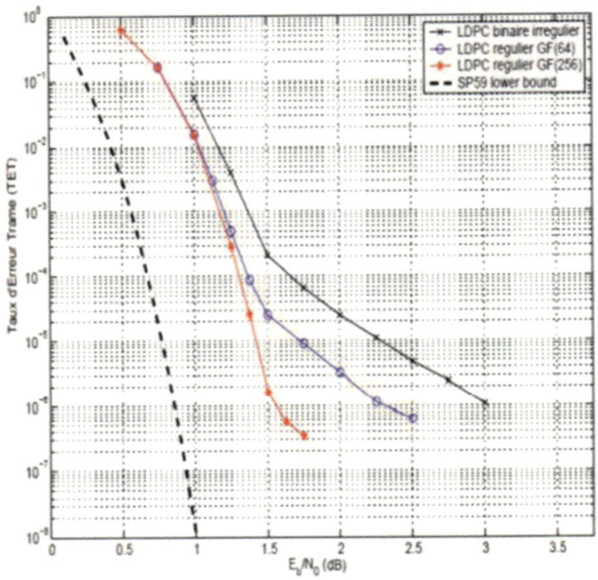

Figure 1.
Binary vs. nonbinary LDPC, N_b = 3008 bits and R = 1/2 [12].

specific type called as characteristic-2 fields are the fields when $n = 2$. All the elements of a characteristic-2 field can be shown in a polynomial format [10].

In coding applications, for $p \leq 32$, it is normal to represent an entire polynomial in $GF(2^p)$ as a single integer value in which individual bits of the integer represent the coefficients of the polynomial. The least significant bit of the integer represents the a_0 coefficient. For example, the polynomial form of the Galois field with 16 elements (known as *GF (16)*, so that $p = 4$), is:

$$a_3x^3 + a_2x^2 + a_1x^1 + a_0x^0$$

with $a_3a_2a_1a_0$ corresponding to the binary numbers *0000* to *1111*.

For any finite field $GF(2^p)$, there exists a primitive polynomial of degree p over $GF(q)$ [11].

Table 2 lists the primitive polynomials.

The following MATLAB functions provide default primitive polynomials for Galois field:

The row vector that supplies the coefficients of the default primitive polynomial for $GF(p^m)$, given by gfprimdf(m,p), is shown in polynomial format by the gfpretty function

For binary field, $p = 1$, while for $p \geq 2$, it represents a nonbinary field. Hence, nonbinary LDPC codes can be visualised as a direct generalisation of binary LDPC codes. **Table 3** shows the example for $p = 3$ while considering the primitive polynomial $1 + x + x^2$.

The binary coefficients of the polynomial representation can be used to represent the Parity Check Matrix of a NB-LDPC code in a binary matrix form (Refer **Figures 2** and **3**).

The field's nonbinary elements can be represented as a polynomial [11] when the field order of an NB-LDPC code is a power of 2, giving the field elements a binary representation.

Consider a normal $(M \times N)$ NB-LDPC code written in a Galois field $GF(q)$, where $q = 2^p$ is the field's order. (d_v, d_c) are the degrees of connection of the variable and check nodes, respectively. The nonzero elements of the NB parity check matrix H associated with the code correspond to the Galois field $GF(2^p = q)$.

Number of primitive polynomials of degree *N*	
N	**No**
1	1
2	1
4	2
8	16
16	2048
32	67,108,864
Table of primitive polynomials up to degree 31	
N	**Primitive polyn omials**
1, 2, 3, 4, 6, 7, 15, 22	$1 + X + X^n$
5, 11, 21, 29	$1 + X^2 + X^n$
10, 17, 20, 25, 28, 31	$1 + X^3 + X^n$
9	$1 + X^4 + X^n$
23	$1 + X^5 + X^n$
18	$1 + X^7 + X^n$
8	$1 + X^2 + X^3 + X^4 + X^n$
12	$1 + X + X^3 + X^4 + X^n$
13	$1 + X + X^4 + X^6 + X^n$
14, 16	$1 + X + X^3 + X^4 + X^n$

Table 2.
Primitive polynomials.

Binary	**Polynomial**	
0 0 0	$0x^2 + 0x + 0$	0
0 0 1	$0x^2 + 0x + 1$	1
0 1 0	$0x^2 + 1x + 0$	x
0 1 1	$0x^2 + 1x + 1$	x + 1
1 0 0	$1x^2 + 0x + 0$	x^2
1 0 1	$1x^2 + 0x + 1$	$x^2 + 1$
1 1 0	$1x^2 + 1x + 0$	$x^2 + x$
1 1 1	$1x^2 + 1x + 1$	$x^2 + x + 1$

Table 3.
For p = 3 *while considering the primitive polynomial* $1 + x + x^2$.

$$H = \begin{bmatrix} h_{00} & h_{01} & h_{02} & h_{03} & 0 & 0 & 0 & 0 & 0 & 0 \\ 0 & h_{10} & 0 & 0 & 0 & h_{11} & 0 & 0 & h_{12} & h_{13} \\ 0 & 0 & h_{20} & 0 & h_{21} & 0 & 0 & h & h_{22} & 0_{23} \\ h_{30} & 0 & 0 & h_{31} & h_{32} & 0 & h_{33} & 0 & 0 & 0 \\ 0 & 0 & 0 & 0 & 0 & h_{40} & h_{41} & h_{42} & 0 & h_{43} \end{bmatrix}$$

Figure 2.
Nonbinary LDPC *parity check matrix [12].*

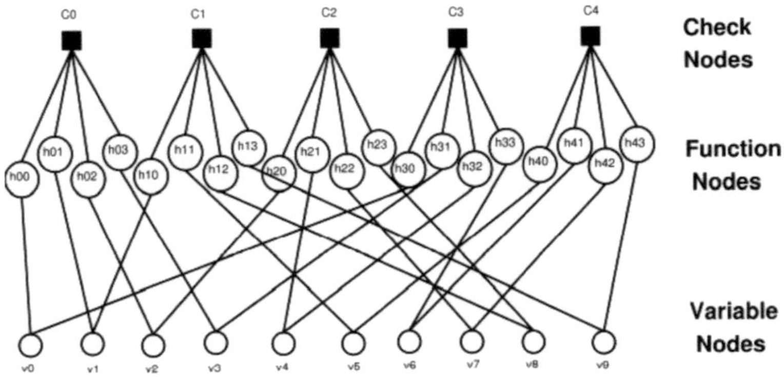

Figure 3.
Tanner graph for a nonbinary parity check matrix [12].

We define a primitive polynomial of degree p for the field:

$$p(x) = a_0 + \ a_1x + \ a_2x^2 \dots + x^p$$

A matrix A of size p × p is also associated to the primitive polynomial [10].

$$A = \begin{bmatrix} 0 & 1 & 0 & 0 & \dots & 0 \\ 0 & 0 & 1 & 0 & \dots & 0 \\ 0 & 0 & 0 & 1 & \dots & 0 \\ . & . & . & . & \dots & . \\ 0 & . & . & . & \dots & 1 \\ a_0 & a_1 & a_2 & a_3 & \dots & a_{p-1} \end{bmatrix}$$

where $[a_0, a_1 \ \dots, a_{p-1}]$ are primitive polynomial p(x)'s coefficients.

Because p(x) represents the field's primitive polynomial, A can be called as its primitive element in matrix representation. As a result, the binary matrix representations of all the other elements of the field are generated by the powers of the matrix A. Thus, the nonzero elements h_{ij} of the parity check matrix can be written in the form of a $(p \times p)$ binary matrices H_{ij}, where H_{ij} is the result of the tranpose of a power of the primitive matrix of the Galois field. Subsequently, the $(M \times N)$ nonbinary parity check matrix can be written in the form of a M_bXN_b binary parity check matrix, where $M_b = pM$ and $N_b = pN$ as shown in **Figure 4**. The zero elements of the parity check matrix are represented with all-zero matrices of size $(p \times p)$.

Modulo-2 addition and multiplication of polynomial representations are used to do arithmetic on the GF(q) elements. Modulo-2 arithmetic over the matrix representations can also be used to do arithmetic on the matrix representations. As a result, in the vectorial domain, the parity check equation may be represented as:

$$\sum_{j:H_{ij} \neq 0} H_{ij} X_j^T = 0^T \tag{6}$$

where H_{ij} is the matrix representation of the Galois field element h_{ij}, X_j is the p-bits binary mapping of the symbol c_j, and *0* is the all zero-component vector.

The methods for decoding binary LDPC codes may be generalised to nonbinary LDPC codes defined over finite fields by performing modifications in

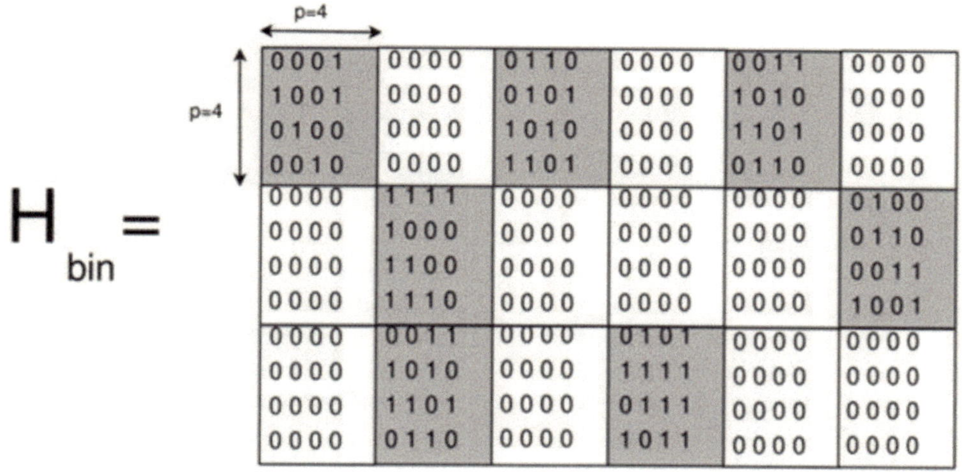

Figure 4.
Nonbinary parity check matrix [12].

correspondence to finite fields. MacKay et al. [13] expanded the belief propagation technique to nonbinary LDPC codes constructed over finite fields. The main obstacle in the development of the hardware realisation of the BP decoding algorithm is its computational complexity, the major factor being the check nodes processing, which is composed of a high number of additions and multiplications. A combination of iterative and list decoding algorithms [12] can be used to design low complexity nonbinary LDPC decoders.

3. Conclusion

In both binary (Galois field) and nonbinary fields, this chapter introduces monic, irreducible, primal polynomials. With examples, the MATLAB functions related to primitive polynomials were also discussed. The relevance of polynomials and the Galois field in creating the nonbinary LDPC code's parity check matrix for error detection and correction has also been described.

Author details

Mani Shankar Prasad and Shivani Verma*
Amity Institute of Space Science and Technology, Amity University Uttar Pradesh, India

*Address all correspondence to: sverma2@amity.edu

References

[1] http://www.math.ucsd.edu/~jmckerna/Teaching/15-16/Spring/103B/l_17.pdf

[2] http://maths.du.ac.in/Events/IWM/talks/S.K.Khanduja.pdf

[3] Chebolu SK, Miná J. Counting irreducible polynomials over finite fields using the inclusion-exclusion principle. Mathematics Magazine. 2011; **84**:369-371. DOI: 10.4169/math.mag.84.5.369 c Mathematical Association of America

[4] https://in.mathworks.com/help/

[5] Hanzlík P. Steganography in Reed-Solomon Codes. Sweden: Luleå University of Technology; 2011

[6] MacKay DJC, Neal RM. Near shannon limit performance of low-density parity check codes. Electronics Letters. 1997;**33**(6):457-458. DOI: 10.1049/el:19970362

[7] MacKay DJC, Davey MC. Evaluation of gallager codes for short block length and high-rate applications. In: Marcus B, Rosenthal J, editors. Codes, Systems, and Graphical Models. The IMA Volumes in Mathematics and its Applications, Vol. 123. New York, NY: Springer; 2001. DOI: 10.1007/978-1-4613-0165-3_6

[8] Ma L, Wang L, Zhang J. Performance Advantage of Non-binary LDPC Codes At High Code Rate under AWGN Channel. 2006 International Conference on Communication Technology. 2006. pp. 1-4. DOI: 10.1109/ICCT.2006.341827

[9] Li G, Fair IJ, Krzymien W. Density evolution for nonbinary LDPC codes under Gaussian approximation. IEEE Transactions on Information Theory. 2009;**55**(3):997-1015

[10] Poulliat C, Fossorier M, Declercq D. Design of non binary ldpc codes using their binary image: algebraic properties. Seattle, USA: In Proceedings of IEEE ISIT; 2006

[11] Poulliat C, Fossorier M, Declercq D. Opitmization of non-binary ldpc codes using their binary images. Munich, Germnay: In Proceedings of IEEE Int. Symp. on Turbo codes; 2006

[12] Shams B. Next generation non-binary LDPC codes. Information Theory [cs.IT]. French: University of Cergy Pontoise; 2010

[13] Davey M, Mackay DJC. Low Density Parity Check Codes Over gf(q). IEEE Communication Letter. 1998;**2**:165-167

Chapter 5

On the Use of Homogeneous Polynomial Yield Functions in Sheet Metal Forming Analysis

Mehmet Firat, Bora Şener, Toros Arda Akşen and Emre Esener

Abstract

Sheet metal forming techniques are a major class of stamping and manufacturing processes of numerous parts such as doors, hoods, and fenders in the automotive and related supplier industries. Due to series of rolling processes employed in the sheet production phase, automotive sheet metals, typically, exhibit a significant variation in the mechanical properties especially in strength and an accurate description of their so-called plastic anisotropy and deformation behaviors are essential in the stamping process and methods engineering studies. One key gradient of any engineering plasticity modeling is to use an anisotropic yield criterion to be employed in an industrial content. In literature, several orthotropic yield functions have been proposed for these objectives and usually contain complex and nonlinear formulations leading to several difficulties in obtaining positive and convex functions. In recent years, homogenous polynomial type yield functions have taken a special attention due to their simple, flexible, and generalizable structure. Furthermore, the calculation of their first and second derivatives are quite straightforward, and this provides an important advantage in the implementation of these models into a finite element (FE) software. Therefore, this study focuses on the plasticity descriptions of homogeneous second, fourth and sixth order polynomials and the FE implementation of these yield functions. Finally, their performance in FE simulation of sheet metal cup drawing processes are presented in detail.

Keywords: Homogeneous polynomials, yield criteria, finite element, plastic anisotropy, cup drawing

1. Introduction

Sheet materials represent significant anisotropic behavior due to their thermomechanical process history. Anisotropy states the variation of the mechanical properties with direction. This material property is determined from tensile test and it is calculated by dividing width plastic strain increments to thickness plastic strain increments. From this definition, it is seen that anisotropy indicates the resistance to the thinning. Therefore, it can be said that increasing anisotropy values improves the deep drawability of the material. Two approaches are applied in the description of the anisotropy. The first approach is the phenomenological approach in which global material behavior is determined according to the average behavior of all grains. The second approach is crystal plasticity which investigates the behavior of one grain to determine the material behavior.

In the phenomenological plasticity approach, the transition from the elastic deformation to plastic deformation is defined with yield functions [1]. A yield function establishes the relationship between principal stresses and yield stress of the material. Plastic flow occurs when the yield function reaches a critical value which is the yield stress of the material. Therefore, yield condition actually indicates a state of equilibrium and it can be defined by the following equation:

$$F = f_y(\overline{\sigma}) - \sigma_y = 0 \tag{1}$$

where $\overline{\sigma}$ and σ_y denote the equivalent and yield stresses, respectively. Eq. (1) defines a surface in three dimensional stress space and it is called as yield surface. According to Drucker's postulate [2], this surface must be closed, convex, and smooth in order to establish a relationship between plastic strain increments and stresses. In the literature, Tresca and von Mises are well known and have been most commonly used yield criteria. However, these yield criteria are isotropic and they could not give satisfactory results for sheet metal forming processes. Therefore, the usage of anisotropic yield functions is required for representation of sheet metal behavior and several anisotropic yield functions have been proposed by researchers. The first phenomenological anisotropic yield function was proposed by Hill in 1948 [3]. Hill added some coefficients to von Mises criterion to transform isotropic von Mises criterion into an anisotropic form. Hill's quadratic criterion could be used for both plane stress (2D) and general stress (3D) states. The criterion has four coefficients for 2D stress state, and it has six coefficients for 3D stress state. These coefficients could be obtained analytically according to stress or plastic strain ratios. Hill48 quadratic criterion has a simple form and useful coefficient identification procedure. However, this criterion could not simultaneously predict the variations of the stress and strain ratios within the sheet plane. Therefore, it could not successfully define the plastic behavior of highly anisotropic materials such as Al-Mg alloys, Ti alloys, etc. Different type yield criteria have been applied to accurately describe the anisotropic behavior of these materials. The most popular approach used to derive an anisotropic yield criterion is the linear transformation method. In this method, Cauchy stress tensor or the deviatoric stress tensor is transformed linearly, and an anisotropic yield function is obtained by substitution of this transformed tensor in an isotropic yield function [4]. Yld89 is one of the functions developed by this approach. Barlat and Lian [5] applied linear transformation method to Hosford 1972 [6] isotropic yield criterion and developed this anisotropic material model. The criterion has four coefficients and it could be used for only 2D stress state. Then, Barlat et al. [7] extended this yield criterion for 3D stress state and developed a criterion has six coefficients in 1991. However, these yield criteria could not accurately describe the anisotropic behavior of especially Al-Mg alloys. Another yield criterion based on linear transformation approach was developed by Karafillis and Boyce [8] in 1993. Karafillis and Boyce generalized Hosford's yield function and proposed an isotropic yield function. Then researchers applied to linear transformation approach and developed an anisotropic yield criterion. They applied their developed yield criterion for modeling of AA2008-T4 alloy and could successfully define the angular variations of both stress and plastic strain ratios of the material. Barlat et al. have inspired by this method and developed Yld2000 and Yld2004 yield criteria, respectively [9, 10]. From these models, Yld2000 could only be used for plane stress condition, whereas the other could be used for both plane stress and general stress states. Yld2004 criterion has 18 coefficients and it could successfully describe in-plane variations of plastic properties of highly anisotropic aluminum alloys. These models are effective in the representation of the anisotropic behavior. However, their parameter identification procedures consist of complex nonlinear formulas and computation of the derivatives is difficult.

Another method which is applied to derive anisotropic yield function is the polynomial approach. Due to inability of quadratic Hill48 criterion, Hill suggested that the usage of general homogeneous polynomials as yield functions in 1950 [11]. In the literature, firstly Gotoh [12, 13] applied this method and modeled the anisotropic behavior of commercial Al-killed steel and Cu-(1/4)H sheets with fourth-order polynomial yield function. Gotoh determined explicitly the coefficients of the polynomial function for these materials and successfully predicted the angular variations of the plastic properties. However, Gotoh did not take into account the convexity of the yield surface in the parameter identification. This deficiency was noticed by Soare et al. [14] and they proposed changes to Gotoh's identification procedure. This modification has contributed to the applicability of the polynomial criteria and important results have been obtained.

In the present work, polynomial yield criteria, their modeling capability and applications on the sheet metal forming simulations have been investigated. Article consists of four sections. In Section 2, the theoretical background of the developed polynomial yield functions are briefly explained. Then, applications of polynomial criteria and results are presented. In Section 4, the main conclusions and findings are summarized.

2. Homogeneous polynomial yield functions

It is seen from the literature that the second, the fourth, and the sixth-order homogeneous polynomials have been used as yield functions. Therefore, the general formulation of these functions are explained in this section.

2.1 Second-order polynomial yield function

Conventional quadratic Hill48 yield criterion can be defined as second-order polynomial yield function (P_2). The form of the criterion for plane stress state could be written as follows:

$$P_2 = a_1{\sigma_x}^2 + a_2{\sigma_y}^2 - 2a_3\sigma_x\sigma_y + 2a_4{\sigma_{xy}}^2 \tag{2}$$

a_1, a_2, a_3, and a_4 are function parameters and they can be determined based on stress or plastic strain ratios. The equations related to stress and plastic strain ratios are given in below. The coefficients determined with stress or strain based definition are distinguished by subscripts σ and R, respectively.

$$a_{1\_\sigma} = 1; a_{2\_\sigma} = \left(\frac{1}{\overline{\sigma}_{90}}\right)^2; a_{3\_\sigma} = \frac{1}{2}\left(1 + \left(\frac{1}{\overline{\sigma}_{90}}\right)^2 - \left(\frac{1}{\overline{\sigma}_b}\right)^2\right); a_{4\_\sigma} = 2\left(\frac{1}{\overline{\sigma}_{45}}\right)^2 - \frac{1}{2}\left(\frac{1}{\overline{\sigma}_b}\right)^2 \tag{3}$$

$$a_{1\_R} = 1; a_{2\_R} = \frac{r_0(1+r_{90})}{r_{90}(1+r_0)}; a_{3\_R} = \frac{r_0}{1+r_0}; a_{4\_R} = \frac{(r_0+r_{90})(1+2r_{45})}{2r_{90}(1+r_0)} \tag{4}$$

2.2 Fourth-order polynomial yield function

For plane stress state, the fourth-order polynomial yield function (P_4) is expressed as following:

$$P_4 = a_1\sigma_x{}^4 + a_2\sigma_x{}^3\sigma_y + a_3\sigma_x{}^2\sigma_y{}^2 + a_4\sigma_x\sigma_y{}^3 + a_5\sigma_y{}^4 + \left(a_6\sigma_x{}^2 + a_7\sigma_x\sigma_y + a_8\sigma_y{}^2\right)\sigma_{xy}{}^2 + a_9\sigma_{xy}{}^4 \tag{5}$$

where a_1, a_2, a_3 .... .a_9 are the material coefficients. In order to determine these nine coefficients, nine experimental data are required. Direct approach for coefficient determination can lead to oscillations in the predictions of the plastic strain or yield stress ratios. Therefore, Soare et al. [14] proposed a different coefficient identification procedure and derived upper and lower bounds on coefficients to obtain a convex and smooth yield surface. In this section, the coefficient identification procedure developed by Soare is explained:

(i) Firstly, the first five coefficients are determined with explicit formulas are given below:

$$a_1 = 1, a_2 = -4r_0/(1+r_0), a_5 = 1/(\overline{\sigma}_{90})^4, a_4 = -4a_5r_{90}/(1+r_{90}) \tag{6}$$

where r_0 and r_{90} indicate plastic strain ratios (r-values) along rolling and transverse directions, whereas $\overline{\sigma}_{90}$ denotes yield stress ratio along transverse direction.

(ii) The coefficient a_3 is determined according to the Eq. (7).

$$a_3 = \left(1/\overline{\sigma}_b{}^4\right) - (a_1 + a_2 + a_4 + a_5) \tag{7}$$

where $\overline{\sigma}_b$ indicates the biaxial yield stress ratio.

(iii) The coefficient a_9 is determined according to Eq. (8)

$$a_9 = \frac{(2/\overline{\sigma}_{45})^4 r_{45}}{1+r_{45}} + \left(1/\overline{\sigma}_b{}^4\right) \tag{8}$$

where $\overline{\sigma}_{45}$ and r_{45} indicate the yield stress and plastic strain ratios along the diagonal direction.

(iv) The coefficients a_6 and a_8 are determined with the minimization of the error (distance) function given in Eq. (9).

$$E = w_1\sum_{i=1}^{2}\left[\frac{(\overline{\sigma}_\theta)_{pred} - (\overline{\sigma}_\theta)_{exp}}{(\overline{\sigma}_\theta)_{exp}}\right]^2 + w_2\sum_{i=1}^{2}\left[\frac{(r_\theta)_{pred} - (r_\theta)_{exp}}{(r_\theta)_{exp}}\right]^2 \tag{9}$$

where w_1 and w_2 are the weight coefficients for stress and plastic strain ratios at the interval angles. In this minimization problem, interval angles could be 15^0-75^0, 30^0-60^0 or 22.5^0-67.5^0. After determination of the coefficients a_6 and a_8, these coefficients are checked for positivity and convexity of the yield surface. In order to obtain convex and smooth yield surface, a_6 and a_8 must satisfy the following inequalities:

$$0 \le a_6 \le 6\sqrt{a_1a_9}, 0 \le a_8 \le 6\sqrt{a_5a_9} \tag{10}$$

v) The coefficient a_7 is determined with Eq. (11)

$$a_7 = \frac{(2/\overline{\sigma}_{45})^4}{1+r_{45}} - 2\left(1/\overline{\sigma}_b{}^4\right) \tag{11}$$

Inequalities related to convexity and positivity conditions are given detailed in [14].

2.3 The sixth-order polynomial yield function

The sixth-order polynomial yield function (P_6) has 16 coefficients for plane stress state and the form of the criterion is given below:

$$\begin{aligned} P_6 = {} & a_1\sigma_x^{\,6} + a_2\sigma_x^{\,5}\sigma_y + a_3\sigma_x^{\,4}\sigma_y^{\,2} + a_4\sigma_x^{\,3}\sigma_y^{\,3} + a_5\sigma_x^{\,2}\sigma_y^{\,4} + a_6\sigma_x\sigma_y^{\,5} + a_7\sigma_y^{\,6} \\ & + \left(a_8\sigma_x^{\,4} + a_9\sigma_x^{\,3}\sigma_y + a_{10}\sigma_x^{\,2}\sigma_y^{\,2} + a_{11}\sigma_x\sigma_y^{\,3} + a_{12}\sigma_y^{\,4}\right)\sigma_{xy}^{\,2} \\ & + \left(a_{13}\sigma_x^{\,2} + a_{14}\sigma_x\sigma_y + a_{15}\sigma_y^{\,2}\right)\sigma_{xy}^{\,4} + a_{16}\sigma_{xy}^{\,6} \end{aligned} \tag{12}$$

The coefficients a_1, a_2, a_6, and a_7 are calculated explicitly and the equations are given below:

$$a_1 = 1, a_2 = -\frac{6r_0}{(1+r_0)}, a_7 = (1/\bar{\sigma}_{90})^6, \quad a_6 = -6r_{90}a_7/(1+r_{90}) \tag{13}$$

The remained coefficients are determined by minimization of the error function given in Eq. (8).

3. Applications of polynomial yield functions

Three validation studies are generally performed in the literature in order to evaluate the prediction capability of orthotropic yield criteria: These are the description of the planar variations of plastic properties, the prediction of the earing profile and number of ears in cup drawing test, and prediction of the thickness strain distributions along the different directions in a drawn part, respectively. Obtained results with polynomial yield functions are presented in below.

3.1 Description of the directional properties

Soare et al. [14] investigated the prediction capability of the polynomial yield functions. They described the anisotropic behavior of AA2090-T3 with P_4 and P_6 yield criteria. **Figures 1** and **2** show the P_4 and P_6 predictions of the angular variation of plastic properties for AA2090-T3 alloy, respectively.

It is seen from **Figures 1** and **2** that both criteria could simultaneously predict the angular variations of stress and plastic strain ratio. In addition to that the predictions of P_6 criterion were more successful than P4 criterion especially at interval

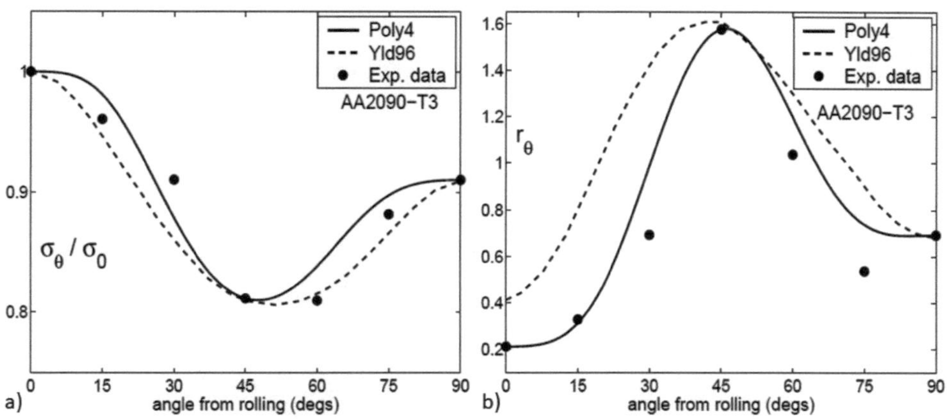

Figure 1.
Comparison of the predicted results from P_4 criterion with experiment (a) stress ratio, (b) r-value.

angles. Sener et al. [15] investigated the evolution of anisotropic behavior of Al5754 with P_2 and P_4 yield criteria. They determined the coefficients of the yield functions at four different plastic strain levels and predicted the angular variations of yield stress and plastic strain ratios. Then, researchers compared the predicted results from yield criteria with experimental data for each plastic strain level. **Figures 3** and **4** show the comparison results for P_2 and P_4 criteria, respectively.

It is seen from **Figures 3** and **4** that P_2 criterion could only accurately predict the variation of r-values in the sheet plane, while P_4 criterion could predict both the angular variations of stress and strain ratios. This result is related to the identification procedures of the yield criteria. As it is declared in Section 2 that, P_2 criterion takes as input either stress or strain ratios. However, the coefficients of P_4 criterion

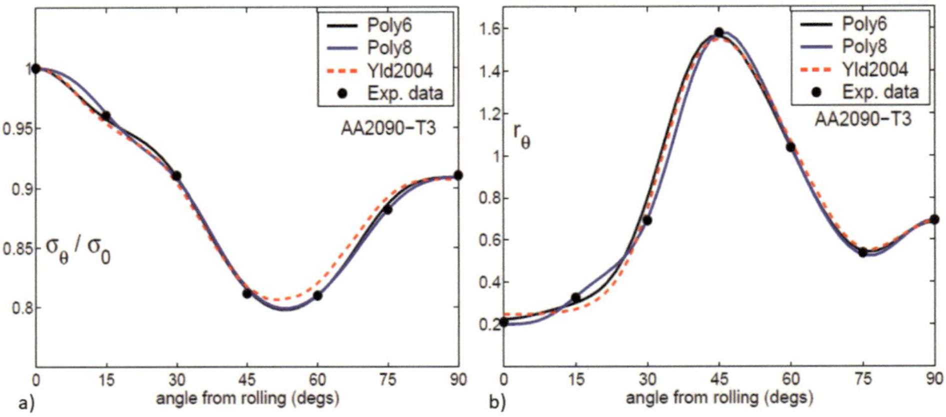

Figure 2.
Comparison of the predicted results from P_6 criterion with experiment (a) stress ratio, (b) r-value.

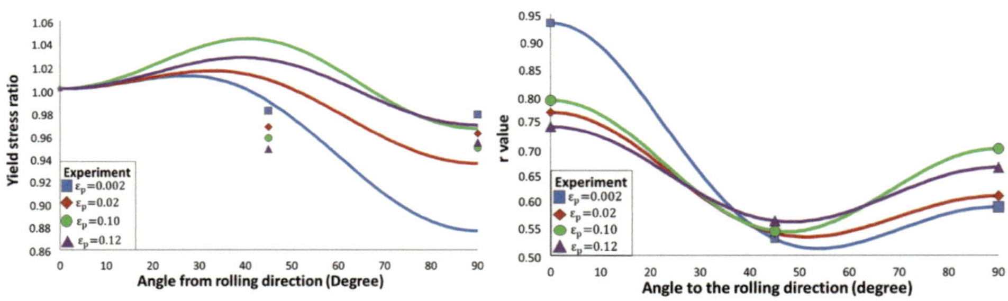

Figure 3.
Comparison of the predicted results from P_2 criterion with experiment (a) stress ratio, (b) r-values.

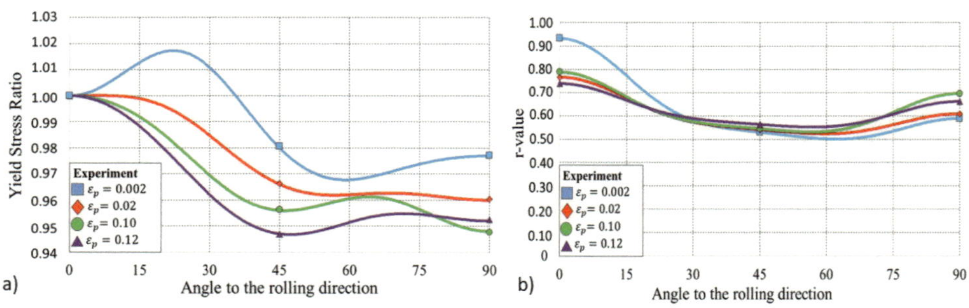

Figure 4.
Comparison of the predicted results from P_4 criterion with experiment (a) stress ratio, (b) r-values.

are calibrated with both stress and strain ratios. In addition to description of the planar anisotropy, researchers investigated the variation of the yield locus shape with plastic strain. **Figure 5a** and **b** show the variation of yield locus contours with plastic strain for P_2 and P_4 yield criteria, respectively.

It is seen from **Figure 5** that the contours of the yield locus are changed with plastic strain and this evolution is more pronounced in P_4 criterion.

3.2 Prediction of the earing profile

Cup drawing is a test which is used for validation of an anisotropic yield criterion. If material has a strong anisotropy, the height of the formed cup is not uniform and a series of crests and valleys are observed around the cup perimeter. This waviness in the top edge of a cup is called as earing and four, six or eight ears could be occurred in a drawn cup depend on the degree of the anisotropy [16, 17]. Soare et al. [14] investigated the prediction capability of polynomial yield functions on the cup drawing test. They implemented P_4 and P_6 yield criteria into FE code ABAQUS and performed FE analyses of the test. Researchers also studied the effect of element type on the predictions and they carried out simulations with shell and solid elements. After FE analyses, they predicted the number of ears, cup height, and compared the numerical results with the Yld96 criterion and experiment. Yld96 criterion was selected as reference by the researchers due to involving the same number of material coefficients of both criteria. **Figures 6** and **7** show the geometry of the drawn cup and the comparison of the predicted cup profiles from P_4 and Yld96 yield criteria with experiment for AA2090-T3 alloy.

It is seen from **Figure 7** that P_4 and Yld96 criteria could successfully predict cup heights, however the predictions of P_4 were closer to the experiment in the rolling direction. Both criteria predicted two extra ears along the transverse direction (90° and 270°). It was also observed that there are no significant differences between the predictions of P_4-2D, and P_4-3D models. Researchers also investigated the capability of P_6 criterion on earing prediction and compared the predictions with Yld2004 and experiment. These comparisons are shown in **Figure 8**.

From the comparisons, it is observed that P_6 criterion could accurately predict both the number of ears and cup height. Another observation in this study is related to Yld2004 and P_6 predictions. Both criteria gave similar results and this shows that P_6 has higher capability in the modeling of the anisotropy.

3.3 Prediction of thickness strains in rectangular cup drawing

Another study related to polynomial yield functions was carried out by Sener et al. [18]. They investigated the anisotropic behavior of AISI 304 stainless steel

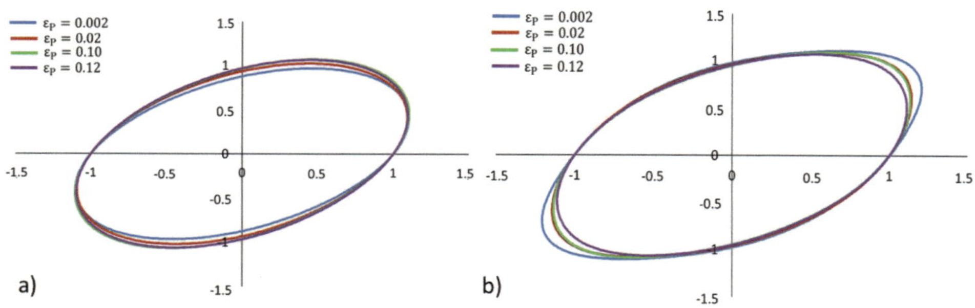

Figure 5.
Variation of the yield locus contours with plastic strain (a) P_2, (b) P_4.

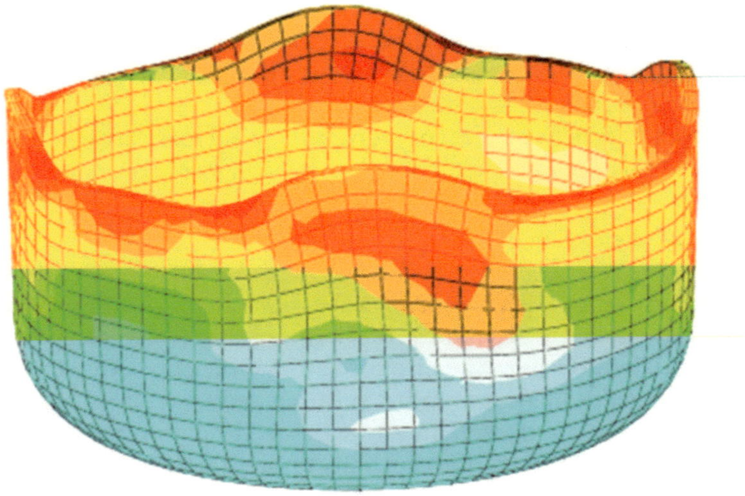

Figure 6.
Drawn cup [14].

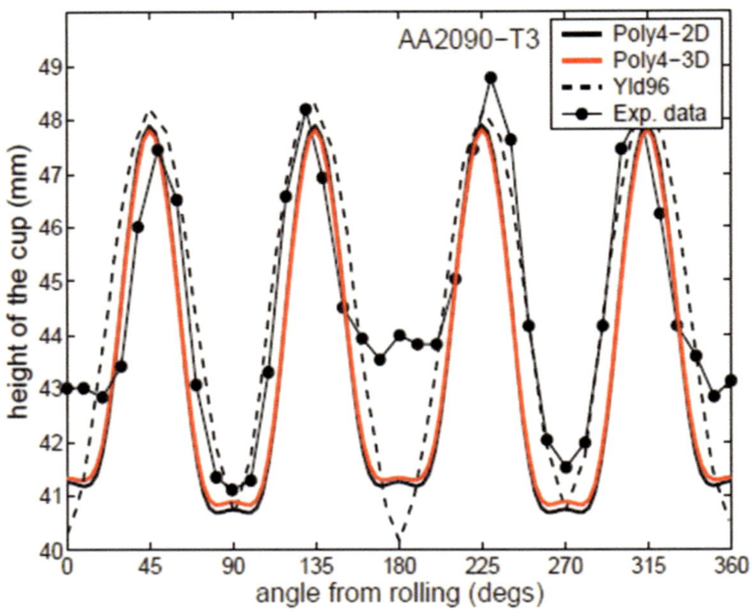

Figure 7.
Experimental and predicted cup profiles from the fourth-order polynomial and Yld96 criteria for AA2090-T3 [14].

with P_4 yield criterion. Investigation was conducted on the uniaxial tensile test and a rectangular cup drawing process. Criterion could successfully describe stress anisotropy and r-value variations. Researchers implemented the criterion into explicit FE code Ls-Dyna by using user defined material subroutines and performed FE simulation of rectangular cup drawing process. They investigated the thickness distributions and flange geometry. **Figures 9** and **10** show the comparisons of the numerical and experimental results in terms of the thickness distributions and flange geometry of the cup.

It is seen from the **Figures 9** and **10** that the predicted thickness distributions and flange geometry matches well with the experimental results. Then, Sener et al. [19] expanded the study [18] and studied the variation of anisotropy during plastic

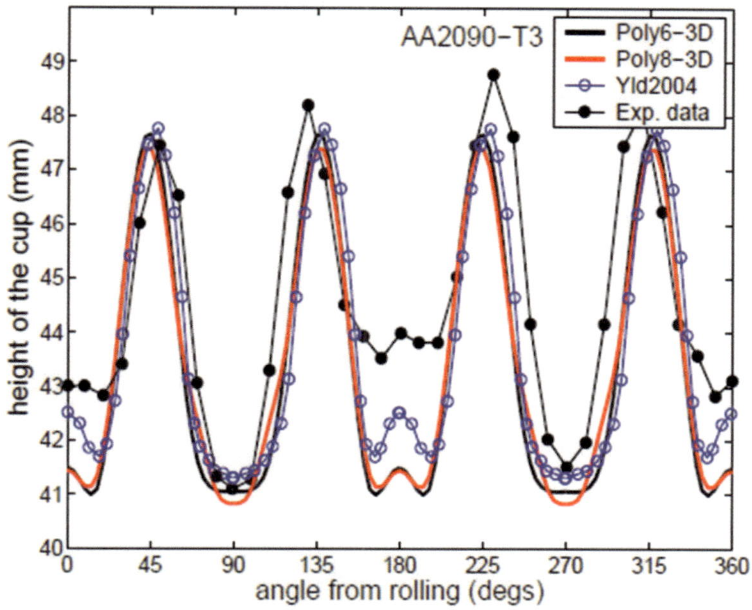

Figure 8.
Experimental and predicted cup profiles from the sixth-eight order polynomial and Yld2004 criteria for AA2090-T3 [14].

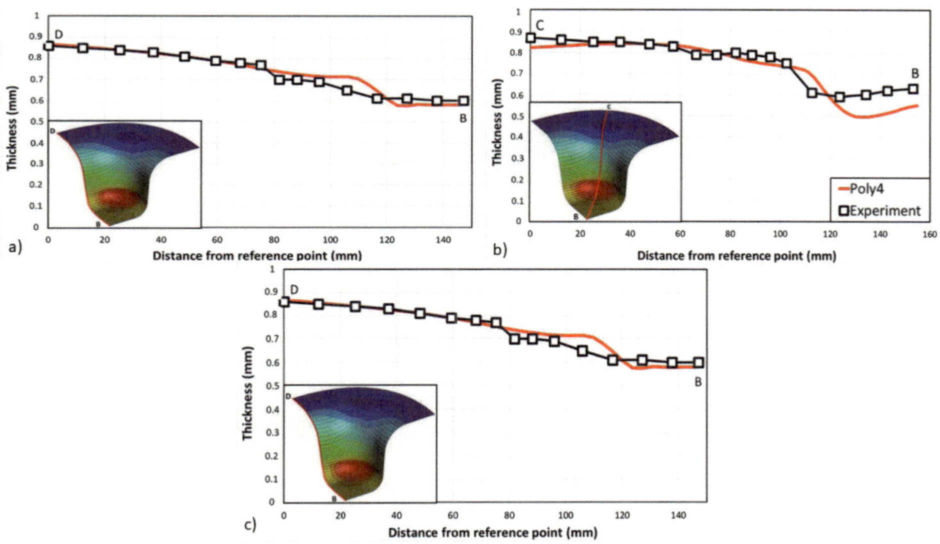

Figure 9.
Numerical and experimental thickness distributions (a) rolling (RD) (b) diagonal (DD), (c) transverse directions (TD).

deformation experimentally and numerically. They carried out FE simulations of same industrial part at different plastic strain levels (0.2%, 2%, 5%, and 18%) and compared P_4 predictions with experimental data. **Figure 11** shows the comparison of the predicted thickness distributions along the three directions with experiment.

It is seen from **Figure 11** that different thickness predictions were obtained at different plastic strain levels. After the comparison of the predicted thickness results with experiment, researchers eliminated two strain levels and then they investigated the flange geometry results (**Figure 12**).

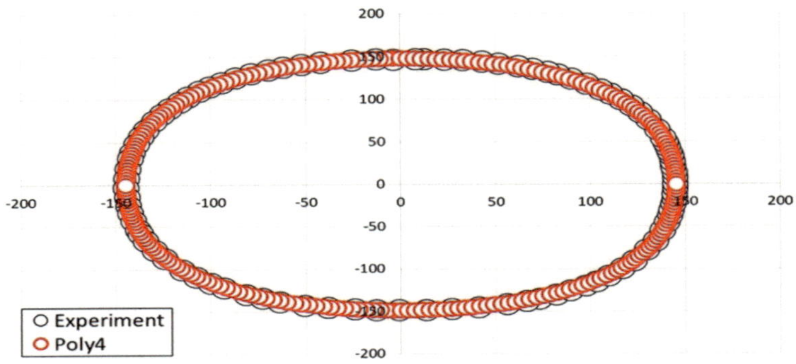

Figure 10.
Numerical and experimental flange geometry.

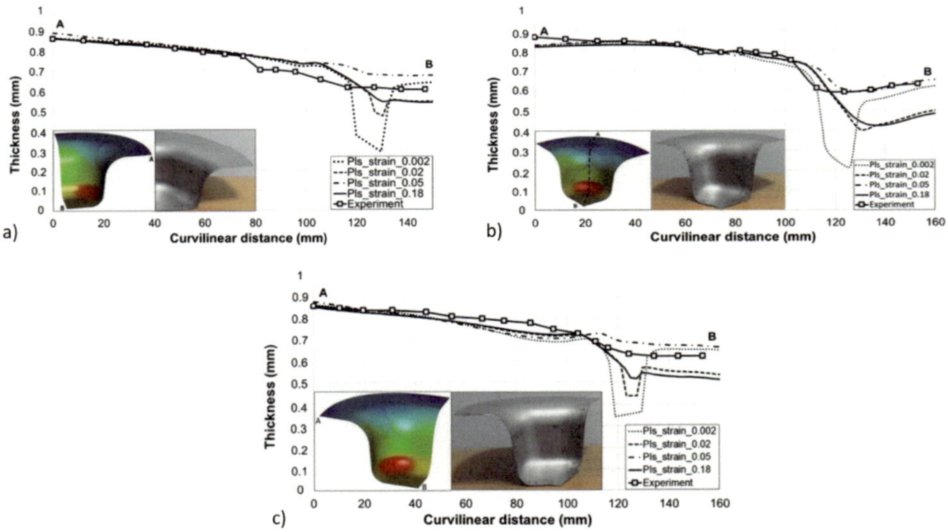

Figure 11.
Comparison of the predicted thickness distributions with experiment (a) RD, (b) DD, (c) TD.

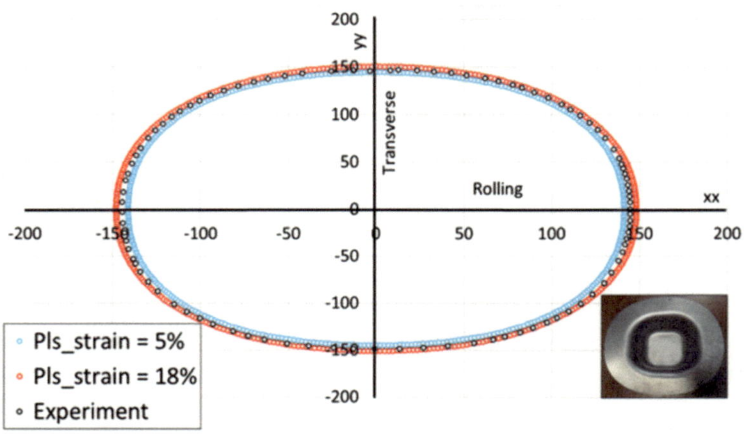

Figure 12.
Comparison of the numerical and experimental flange geometry.

From the comparison of the predicted and experimental flange geometry results, it is seen that numerical results were matched well with the experiment.

4. Conclusions

In the present study, homogeneous anisotropic polynomial yield functions, their types, and application areas in the metal forming process were investigated. In the literature, generally anisotropic yield functions derived from linear transformation approach are used. These functions have high modeling capability and they could be used for different materials. However, yield functions based on linear transformation approach have some disadvantages. They have complex coefficient identification procedure and nonlinear formulas. Therefore, calculations of the first and second order gradients of these models are difficult and it causes to difficulties in the implementation of the models into FE codes. On the other hand, polynomial yield functions have a generalized, simple structure and derivatives of these functions could easily calculated.

It is seen from the studies carried out in the literature that researchers generally use the fourth and the sixth order polynomial functions to model of the anisotropic behavior of the materials. Based on the results obtained from the studies performed in the literature, the following conclusions could be drawn:

a. Homogeneous polynomial yield functions have high modeling capability in the description of anisotropic behavior.

b. Homogeneous polynomial yield functions could be used for both plane stress and generalized stress state. This provides the flexibility to the polynomial yield criteria.

c. Sixth-order homogeneous polynomial yield function could predict six or more ears in a deep drawn cup.

d. Homogeneous polynomial yield functions could model body centered and face centered cubic materials without the need of any exponent related to crystallographic structure.

e. Apart from the linear transformation approach, polynomial models may not satisfy convexity requirements for each stress state. Therefore, the user should consider convexity conditions and has to investigate the model parameters in terms of convexity and positivity conditions.

f. The modeling capability of the fourth-order polynomial yield function is similar with Yld96 yield function, whereas predictions of the sixth-order polynomial yield function close to Yld2004-18p model.

Author details

Mehmet Firat[1*], Bora Şener[2], Toros Arda Akşen[1] and Emre Esener[3]

1 The University of Sakarya, Turkey

2 Yildiz Technical University, Turkey

3 Bilecik Seyh Edebali University, Turkey

*Address all correspondence to: firat@sakarya.edu.tr

References

[1] Banabic D. Sheet Metal Forming Processes. Springer-Verlag: Berlin Heidelberg; 2010. 30 p. DOI: 10.1007/978-3-540-88113-1_1

[2] Drucker DC. A more fundamental approach to plastic stress-strain relations. In: 1 st U.S. Congress of Applied Mechanics (ASME), New York, 1952, p.116-126

[3] Hill R. A theory of the yielding and plastic flow of anisotropic metals. Proceedings of the Royal Society of London Series. 1948; 281-297. DOI: 10.1098/rspa.1948.0045

[4] Barlat F, Yoon JW, Cazacu O. On linear transformations of stress tensors for the description of plastic anisotropy. International Journal of Plasticity. 2007; 23: 876-896

[5] Barlat F, Lian J. Plastic behavior and stretchability of sheet metals. Part I: a yield function for orthotropic sheets under plane stress conditions. International Journal of Plasticity. 1989; 5: 51-66

[6] Hosford WF. A generalized isotropic yield criterion. Journal of Applied Mechanics. 1972; 39: 607-609

[7] Barlat F, Lege DJ, Brem JC. A six-component yield function for anisotropic materials. International Journal of Plasticity. 1991; 7: 693-712

[8] Karafillis AP, Boyce MC. A general anisotropic yield criterion using bounds and a transformation weighting tensor. Journal of the Mechanics and Physics of Solids. 1993; 41: 1859-1886

[9] Barlat F, Brem JC, Yoon JW, Chung K, Dick RE, Lege DJ, Pourboghrat F, Choi S-H, Chu E. Plane stress yield function for aluminum alloy sheets-part I:theory. International Journal of Plasticity. 2003; 19: 1297-1319

[10] Barlat F, Aretz H, Yoon JW, Karabin ME, Brem JC, Dick RE. Linear transformation-based anisotropic yield functions. International Journal of Plasticity. 2005; 21: 1009-1039

[11] Hill R. The Mathematical Theory of Plasticity. 1st ed. Oxford University Press: New York; 1950. 330p.

[12] Gotoh M. A theory of plastic anisotropy based on a yield function of fourth order (plane stress state)-I. International Journal of Mechanical Sciences. 1977; 19: 505-512.

[13] Gotoh M. A theory of plastic anisotropy based on a yield function of fourth order (plane stress state)-II. International Journal of Mechanical Sciences. 1977; 19: 513-520.

[14] Soare S, Yoon JW, Cazacu O. On the use of homogeneous polynomials to develop anisotropic yield functions with applications to sheet forming. International Journal of Plasticity. 2008; 24: 915-944.

[15] Sener B, Esener E, Firat M. Determining the effect of yield criterias on plasticity modeling in different plastic strain levels. In: Proceedings of the International Automotive Technologies Congress (OTEKON 2018); 07-08 May 2018; Bursa: 2018, p.754-760.

[16] Hosford W, Caddell RM. Metal Forming Mechanics and Metallurgy: Cambridge University Press; 2007.228 p.

[17] Yoon JW, Barlat F, Dick RE, Karabin ME. Prediction of six or eight ears in a drawn cup based on a new anisotropic yield function. International Journal of Plasticity. 2006; 22: 174-193.

[18] Sener B, Kilicarslan ES, Firat M. Modelling anisotropic behavior of AISI

304 stainless steel sheet using a fourth-order polynomial yield function. Procedia Manufacturing. 2020; 47: 1456-1461.

[19] Sener B, Esener E, Firat M. Modeling plastic anisotropy evolution of AISI 304 steel sheets by a polynomial yield function. SN Applied Sciences. 2021; 3: 1-12.

Chapter 6

On the Irreducible Factors of a Polynomial and Applications to Extensions of Absolute Values

Lhoussain El Fadil and Mohamed Faris

Abstract

Polynomial factorization over a field is very useful in algebraic number theory, in extensions of valuations, etc. For valued field extensions, the determination of irreducible polynomials was the focus of interest of many authors. In 1850, Eisenstein gave one of the most popular criterion to decide on irreducibility of a polynomial over $\mathbb{Q}$. A criterion which was generalized in 1906 by Dumas. In 2008, R. Brown gave what is known to be the most general version of Eisenstein-Schönemann irreducibility criterion. Thanks to MacLane theory, key polynomials play a key role to extend absolute values. In this chapter, we give a sufficient condition on any monic plynomial to be a key polynomial of an absolute value, an irreducibly criterion will be given, and for any simple algebraic extension $L = K(\alpha)$, we give a method to describe all absolute values of L extending $||$, where $(K, ||)$ is a discrete rank one valued field.

Keywords: Irreducibly criterion, irreducible factors, Extensions of absolute values, Newton polygon's techniques

1. Introduction

Polynomial factorization over a field is very useful in algebraic number theory, for prime ideal factorization. It is also important in extensions of valuations, etc. For valued field extensions, the determination of irreducible polynomials was the focus of interest of many authors (cf. [1–7]). In 1850, Eisenstein gave one of the most popular criterion to decide on irreducibility of a polynomial over $\mathbb{Q}$ [1]. A criterion which was generalized in 1906 by Dumas in [8], who showed that for a polynomial $f(x) = a_nx^n + a_{n-1}x^{n-1} + \dots + a_0 \in \mathbb{Q}[x]$ $(a_0 \neq 0)$, if $\nu_p(a_n) = 0$, $n\nu_p(a_i) \geq (n-i)\nu_p(a_0) > 0$ for every $0 = i, \dots, n-1$, and $gcd(\nu_p(a_0), n) = 1$ for some prime integer p, then $f(x)$ is irreducible over $\mathbb{Q}$. In 2008, R. Brown gave what is known to be the most general version of Eisenstein-Schönemann irreducibility criterion [9]. He showed for a valued field (K, ν) and for a monic polynomial $f(x) = \phi^n(x) + a_{n-1}(x)\phi^{n-1}(x) + \dots + a_0(x) \in R_\nu[x]$, where R_ν is a valuation ring of a discrete rank one valuation and ϕ being a monic polynomial in $R_\nu[x]$ whose reduction $\overline{\phi}$ is irreducible over $\mathbb{F}_\nu$, $a_i(x) \in R_\nu[x]$, $\deg(a_i) < \deg(\phi)$ for every $i = 0, \dots, n-1$, if $\nu(a_i) \leq (1 - i/n)\nu(a_0)$ for every $i = 0, \dots, n-1$ and $\gcd(\nu(a_0), n) = 1$, then $f(x)$ is irreducible over the field K. In this paper, based on absolute value, we give an irreducibility criterion of monic polynomials. More precisely, let $(K, ||)$ be a discrete rank one valued field, $R_{||}$ its valuation ring, $\mathbb{F}_{||}$, its residue field, and $\Gamma = |K^*|$ its value group, we show that for a monic polynomial $f(x) = \phi^n(x) + a_{n-1}(x)\phi^{n-1}(x) + \dots + a_0(x) \in R_{||}[x]$, where ϕ being a monic polynomial in $R_{||}[x]$ whose reduction $\overline{\phi}$ is irreducible over $\mathbb{F}_{||}$, $a_i(x) \in R_{||}[x]$, $\deg(a_i) < \deg(\phi)$ for every $i = 0, \dots, n-1$, if $|a_{n-i}|_\infty \geq \gamma^i$ for every $i = 0, \dots, n-1$ and n is the smallest integer satisfying $\gamma^n \in \Gamma$, where $\gamma = \left(|a_0|_\infty\right)^{1/n}$, then $f(x)$ is irreducible over K. Similarly for the results of extensions of valuations given in [10, 11], for any simple algebraic extension $L = K(\alpha)$, we give a method to describe all absolute values of L extending $||$, where $(K, ||)$ is a discrete rank one valued field. Our results are illustrated by some examples.

2. Preliminaries

2.1 Newton polygons

Let $L = \mathbb{Q}(\alpha)$ be a number field generated by a complex root α of a monic irreducible polynomial $f(x) \in \mathbb{Z}[x]$ and $\mathbb{Z}_L$ the ring of integers of L. In 1894, K. Hensel developed a powerful approach by showing that the prime ideals of $\mathbb{Z}_L$ lying above a prime p are in one–one correspondence with monic irreducible factors of $f(x)$ in $\mathbb{Q}_p[x]$. For every prime ideal corresponding to any irreducible factor in $\mathbb{Q}_p[x]$, the ramification index and the residue degree together are the same as those of the local field defined by the irreducible factors [6]. These results were generalized in ([12], Proposition 8.2). Namely, for a rank one valued field (K, ν), R_ν its valuation ring, and $L = K(\alpha)$ a simple extension generated by $\alpha \in \overline{K}$ a root of a monic irreducible polynomial $f(x) \in R_\nu[x]$, the valuations of L extending ν are in one–one correspondence with monic irreducible factors of $f(x)$ in $K^h[x]$, where K^h is the henselization of $(K, ||)$ will be defined later. So, in order to describe all valuations of L extending ν, one needs to factorize the polynomial $f(x)$ into monic irreducible factors over K^h. The first step of the factorization was based on Hensel's lemma. Unfortunately, the factors provided by Hensel's lemma are not necessarily irreducible over K^h. The Newton polygon techniques could refine the factorization. Namely, theorem of the product, theorem of the polygon, and theorem of residual polynomial say that we can factorize any factor provided by Hensel's lemma, with as many sides of the polygon and with as many of irreducible factors of the residual polynomial. For more details, we refer to [7, 13] for Newton polygons over p-adic numbers and [14, 15] for Newton polygons over rank one discrete valued fields. As our proofs are based on Newton polygon techniques, we recall some fundamental notations and techniques on Newton polygons. Let (K, ν) be a rank one discrete valued field ($\nu(K^*) = \mathbb{Z}$), R_ν its valuation ring, M_ν its maximal ideal, $\mathbb{F}_\nu$ its residue field, and $\left(K^h, \nu^h\right)$ its henselization; the separable closure of K in $\hat{K}$, where $\hat{K}$ is the completion of $(K, ||)$, and $||$ is an associated absolute value of ν. By normalization, we can assume that $\nu(K^*) = \mathbb{Z}$, and so M_ν is a principal ideal of R_ν generated by an element $\pi \in K$ satisfying $\nu(\pi) = 1$. Let also ν be the Gauss's extension of ν to $K^h(x)$. For any monic polynomial $\phi \in R_\nu[x]$ whose reduction modulo M_ν is irreducible in $\mathbb{F}_\nu[x]$, let $\mathbb{F}_\phi$ be the field $\frac{\mathbb{F}_\nu[x]}{(\overline{\phi})}$.

Let $f(x) \in R_\nu[x]$ be a monic polynomial and assume that $\overline{f(x)}$ is a power of $\overline{\phi}$ in $\mathbb{F}_\nu[x]$, with $\phi \in R_\nu[x]$ a monic polynomial, whose reduction is irreducible in $\mathbb{F}_\nu[x]$. Upon the Euclidean division by successive powers of ϕ, we can expand $f(x)$ as follows $f(x) = \sum_{i=0}^{l} a_i(x)\phi(x)^i$, where $\deg(a_i) < \deg(\phi)$ for every $i = 0, \dots, l$. Such a ϕ-expansion is unique and called the ϕ-expansion of $f(x)$. The ϕ-Newton polygon of f, denoted by $N_\phi(f)$ is the lower boundary of the convex envelope of the set of points $\{(i, \nu(a_i)), i = 0, \dots, l\}$ in the Euclidean plane. For every edge S_j, of the polygon $N_\phi(f)$, let l_j be the length of the projection of S_j to the x-axis and H_j the length of its projection to the y-axis. l_j is called the length of S_j and H_j is its height. Let $d_j = \gcd(l_j, H_j)$ be the degree of S_j, $e_j = \frac{l_j}{d_j}$ the ramification degree of S_j, and $-\lambda_j = -\frac{H_j}{l_j} \in \mathbb{Q}$ the slope of S_j. Geometrically, we can remark that $N_\phi(f)$ is the process of joining the obtained edges $S_1, \dots, S_r$ ordered by increasing slopes, which can be expressed by $N_\phi(f) = S_1 + \dots + S_r$. The segments $S_1, \dots,$ and S_r are called the sides of $N_\phi(f)$. The principal ϕ-Newton polygon of $f(x)$, denoted by $N_\phi^+(f)$, is the part of the polygon $N_\phi(f)$, which is determined by joining all sides of negative slopes. For every side S of the polygon $N_\phi^+(f)$ of slope $-\lambda$ and initial point (s, u_s), let l be its length, H its height and e the smallest positive integer satisfying $e\lambda \in \mathbb{Z}$. Since $l\lambda = H \in \mathbb{Z}$, we conclude that e divides l, and so $d = l/e \in \mathbb{Z}$ called the degree of S. Remark that $d = \gcd(l, H)$. For every $i = 0, \dots, l$, we attach the following residue coefficient $c_i \in \mathbb{F}_\phi$:

$$c_i = \begin{cases} 0, & \text{if } (s+i, u_{s+i}) \text{ lies strictly above } S \\ \left(\dfrac{a_{s+i}(x)}{\pi^{u_{s+i}}}\right) \pmod{(\pi, \phi)}, & \text{if } (s+i, u_{s+i}) \text{ lies on } S. \end{cases} \tag{1}$$

where (π, ϕ) is the maximal ideal of $R_\nu[x]$ generated by π and ϕ.

Let $\lambda = -h/e$ be the slope of S, where $h = H/d$ and $d = l/e$. Notice that, the points with integer coordinates lying in S are exactly $(s, u_s), (s+e, u_s - h), \dots, (s+de, u_s - dh)$. Thus, if i is not a multiple of e, then $(s+i, u_{s+i})$ does not lie on S, and so $c_i = 0$. It follows that the candidate abscissas which yield nonzero residue coefficient are $s, s+e, \dots,$ and $s+de$. Let $R_\lambda(f)(y) = t_d y^d + t_{d-1} y^{d-1} + \dots + t_1 y + t_0 \in \mathbb{F}_\phi[y]$ be the residual polynomial of $f(x)$ associated to the side S, where for every $i = 0, \dots, d$, $t_i = c_{ie}$. For every $\lambda \in \mathbb{Q}^+$, the λ-component of $N_\phi(f)$ is the largest segment of $N_\phi(f)$ of slope $-\lambda$. If $N_\phi(f)$ has a side S of slope $-\lambda$, then $T = S$. Otherwise, T is reduced to a single point; the end point of a side S_i, which is also the initial point of S_{i+1} if $\lambda_{i+1} < \lambda < \lambda_i$ or the initial point of $N_\phi(f)$ if $\lambda_i < \lambda$ for every side S_i of $N_\phi(f)$ or the end point of $N_\phi(f)$ if $\lambda_i < \lambda$ for every side S_i of $N_\phi(f)$. In the sequel, we denote by $R_\lambda(f)(y)$, the residual polynomial of $f(x)$ associated to the λ-component of $N_\phi(f)$.

The following are the relevant theorems from Newton polygon. Namely, theorem of the product and theorem of the polygon. For more details, we refer to [15].

Theorem 2.1. (theorem of the product) *Let $f(x) = f_1(x) f_2(x)$ in $R_\nu[x]$ be monic polynomials such that $\overline{f(x)}$ is a positive power of $\overline{\phi}$. Then for every $\lambda \in \mathbb{Q}^+$, if T_i is the λ-componenet of $N_\phi(f_i)$, then $T = T_1 + T_2$ is the λ-componenet of $N_\phi(f)$ and*

$$R_\lambda(f)(y) = R_\lambda(f_1)(y) R_\lambda(f_2)(y)$$

up to multiplication by a nonzero element of $\mathbb{F}_\phi$.

Theorem 2.2. (theorem of the polygon) *Let $f \in R_\nu[x]$ be a monic polynomial such that $\overline{f(x)}$ is a positive power of $\overline{\phi}$. If $N_\phi(f) = S_1 + \dots + S_g$ has g sides of slope $-\lambda_1, \dots, -\lambda_g$ respectively, then we can split $f(x) = f_1 \times \dots \times f_g(x)$ in $K^h[x]$, such that $N_\phi(f_i) = S_i$ and $R_{\lambda_i}(f_i)(y) = R_{\lambda_i}(f)(y)$ up to multiplication by a nonzero.*

Theorem 2.3. (theorem of the residual polynomial) *Let $f \in R_\nu[x]$ be a monic polynomial such that $N_\phi(f) = S$ has a single side of finite slope $-\lambda$. If $R_\lambda(f)(y) = \prod_{i=1}^t \psi_i(y)^{a_i}$ is the factorization in $\mathbb{F}_\phi[y]$, then $f(x)$ splits as $f(x) = f_1(x) \times \cdots \times f_t(x)$ in $K^h[x]$ such that $N_\phi(f_i) = S_i$ has a single side of slope $-\lambda$ and $R_\alpha(f_i)(y) = \psi_i(y)^{a_i}$ up to multiplication by a nonzero element of $\mathbb{F}_\phi$ for every $i = 1, \cdots, t$.*

2.2 Absolute values

Let $||$ be an absolute value of K; a map $||: K \to \mathbb{R}^+$, which satisfies the following three axioms:

1. $|a| = 0$ if and only if $a \neq 0$,

2. $|ab| = |a||b|$, and

3. $|a+b| \leq |a| + |b|$. (triangular inequality)

for every $(a,b) \in K^2$.

If the triangular inequality is replaced by an ultra-inequality, namely $|a+b| \leq \max\{|a|,|b|\}$ for every $(a,b) \in K^2$, then the absolute value $||$ is called a non archimidean absolute value and we say that $(K,||)$ is a non archimidean valued field.

Lemma 2.4. *Let $(K,||)$ be a valued field. Then $||$ is a non archimidean absolute value if and only if the set $\{|n1_K|, n \in \mathbb{N}\}$ is bounded in $\mathbb{R}$.*

Proof. By induction if $||$ is a non archimidean absolute value, then the set $\{|n1_K|, n \in \mathbb{N}\}$ is bounded by 1.

Conversely, assume that there exists $M \in \mathbb{R}^+$ such that $|n1_K| \leq M$ for every $n \in \mathbb{N}$. Let $(a,b) \in K^2$, $n \in \mathbb{N}$, and set $m = sup(|a|,|b|)$. Then $|a+b|^n = |\sum_{k=0}^{n} \binom{n}{k} a^k b^{n-k}|$, where $\binom{n}{k}$ is the binomial coefficient. As $||$ is a non archimidean absolute value, $|a+b|^n \leq \sup\{|\binom{n}{k} 1_K| \cdot |a^k b^{n-k}| \leq M m^n$. Thus $|a+b| \leq M^{1/n} m$. Go over the limit, we obtain $|a+b| \leq m = sup(|a|,|b|)$ as desired. □

Exercices 1. Let $(K,||)$ be a valued field.

1. Show that if K is a finite field, then $||$ is a non archimidean absolute value.

2. More precisely, show that if K is a finite field, then $||$ is trivial; $|x| = 1$ for every $x \in K^*$.

3. Let $\nu : K^* \to \mathbb{R}$ be the map defined by $\nu(a) = -Ln(|a|)$ for every $a \in K^*$, where Ln is the Napierian logarithm defined on $\mathbb{R}^+$. Show that $||$ is a non archimidean absolute value if and only if ν is a valuation of K. ν is called an associated valuation to $||$.

4. Show that if $||$ is a non archimidean absolute value and $(a,b) \in K^2$ such that $|a| \neq |b|$, then $|a+b| = \max(|a|,|b|)$.

5. Let p be a prime integer and $||_p : \mathbb{Q} \to \mathbb{R}^+$, defined by $|a|_p = p^{-\nu_p(a)}$ for every $a \in K^*$, where ν_p is the p-adic valuation on $\mathbb{Q}$. Show that $||_p$ is a non archimidean absolute value of $\mathbb{Q}$.

2.3 Characteristic elements of an absolute value

Let $(K,||)$ be a non archimedian valued field.

Let $R_{||} = \{a \in K, |a| \leq 1\}$ and $M_{||} = \{a \in K, |a| < 1\}$. Then $R_{||}$ is a valuation ring, called the valuation ring of $||$, $M_{||}$ its maximal ideal, and so $\mathbb{F}_{||} = R_{||}/M_{||}$ is a field, called the residue field of $||$.

Exercices 2. Let p be a prime integer and $||$ the p-adic absolute value of $\mathbb{Q}$, defined by $|a| = p^{-\nu_p(a)}$ for every $a \in \mathbb{Z}$, where $\nu_p(a)$ is the greatest integer satisfying $p^{\nu_p(a)}$ divides a for $a \neq 0$ and $\nu_p(0) = \infty$.

1. Show that $(\mathbb{Q}, ||)$ is a non archimidean valued field.

2. Determine the characteristic elements of $||$.

Exercices 3. Let $(K, ||)$ be a non archimidean valued field.

1. Show that $\Gamma = |K^*|$ is a sub-group of $(\mathbb{R}^*, .)$, called the value group of $||$.

2. For every polynomial $A = \sum_{i=0}^{n} a_i x^i \in K[x]$, let $|A|_\infty = \max\{|a_i|, i = 0, \ldots, n\}$. Show that, extended by $|A/B|_\infty = |A|_\infty/|B|_\infty$ for every $(A, B) \in K[x]^2$ with $B \neq 0$, $||_\infty$ define an absolute value on $K(x)$ called the Gauss's extension of $||$.

3. For every polynomial $P = \sum_{i=0}^{n} p_i(x)\phi^i \in K[x]$, let $|P|_\phi = \max\left\{|p_i|_\infty, i = 0, \ldots, n\right\}$. Show that, extended by $|A/B|_\phi = |A|_\phi/|B|_\phi$ for every $(A, B) \in K[x]^2$ with $B \neq 0$, $||_\phi$ define an absolute value on $K(x)$.

2.4 Completion and henselization

Let $(K, ||)$ be a valued field and consider the map $d : K \times K \to \mathbb{R}_{\geq 0}$, defined by $d(a, b) = |a - b|$. Then d is a metric on K.

Definition 1. A sequence $(u_n) \in K^{\mathbb{N}}$ is said to be a Cauchy sequence if for every positive real number ε, there exists an integer N such that for every natural numbers $m, n \geq N$, we have $|u_n - u_m| \leq \varepsilon$.

Example 1.

Any convergente sequence of $(K, ||)$ is a Cauchy sequence.

The converse is false, indeed, it suffices to consider the valued field $(\mathbb{Q}, ||_0)$ with $||_0$ is the usual absolute value of $\mathbb{Q}$ and $u_n = 1 + 1/1! + \ldots + 1/n!$ for every natural integer n. Then (u_n) is a Cauchy sequence, which is not convergente.

Definition 2. A valued field $(K, ||)$ is said to be complete if every Cauchy sequence of $(K, ||)$ is convergente.

Example 2.

1. $(\mathbb{R}, ||_0)$ is a complete valued field.

2. $(\mathbb{Q}, ||_0)$ is not a complete valued field.

Definition 3. Let $(K, ||)$ be a valued field, L/K an extension of fields, and $||_L$ an absolute value of L.

1. We say that $||_L$ extends $||$ if $||_L$ and $||$ coincide on K. In this case $(L, ||_L)/(K, ||)$ is called a valued field extension.

2. Let $(L, ||_L)/(K, ||)$ be a valued field extension and $\Delta = |L^*|_L$. Then $e = |\Delta/\Gamma|$ the cardinal order of Δ/Γ, is called the ramification index of the extension and $f = [\mathbb{F}_{||_L} : F_{||}]$ is called its residue degree.

Definition 4. Let $(K_1, ||_1)$ and $(K_2, ||_2)$ be two valued fields and $f : K_1 \to K_2$ be an isomorphism of fields. f is said to be an isomorphism of valued fields if it preserves the absolute values.

Exercices 4. Let $(L, ||_L)/(K, ||)$ be a valued field extension.

1. Show that if $||$ is a non archimidean absolute value, then $||_L$ is a non archimidean absolute value.

2. Assume that $(K, ||)$ is a non archimidean valued field. Show that the convergence of a series in K is equivalent to the convergence of its general term to 0.

3. Let $\Gamma = |K^*|$ and $\Delta = |L^*|_L$. Show that if Γ is a discrete rank one Abelian group and L/K is a finite extension, then Δ is a discrete rank one Abelian group and $[L : K] = ef$, where e is the ramification index of the extension and f is its residue degree.

Theorem 2.5. ([16], Theorem 1.1.4)
There exists a complete valued field $(L, ||_L)$, *which extends* $(K, ||)$.

Definition 5. The smallest complete valued field extending $(K, ||)$ is called the completion of $(K, ||)$ and denoted by $\hat{K}$.

Furtheremore, the completion is unique up to a valued fields isomorphism.

Now we come to an important property of complete fields. This theorem is widely known as Hensel's Lemma. For the proof, we refer to ([16], Lemma 4.1.3).

Theorem 2.6. (*Hensel's lemma*)
Let $f \in R_{||}[x]$ *be a monic polynomial such that* $\overline{f(x)} = g_1(x)g_2(x)$ *in* $\mathbb{F}_{||}[x]$ *and* $g_1(x)$ *and* $g_2(x)$ *are coprime in* $\mathbb{F}_{||}[x]$. *If* $(K, ||)$ *is a complete valued field valued field, then there exists two monic polynomials* $f_1(x)$ *and* $f_2(x)$ *in* $R_{||}[x]$ *such that* $\bar{f}_1(x) = g_1(x)$ *and* $\bar{f}_2(x) = g_2(x)$.

The following example shows that for any prime integer p, Hensel's lemma is not applicable in $(\mathbb{Q}, ||)$, with $||$ is the p-adic absolute value defined by $|a| = p^{-\nu_p(a)}$. Indeed, let q be a prime integer which is coprime to p, $n \geq 2$ an integer, and $f(x) = x^n + qx + pq \in Z[x]$. First $\bar{f}(x) = x(x^{n-1} + q)$ in $\mathbb{F}_{||}[x]$. As $f(x)$ is q-Eisenstein $f(x)$ is irreducible over $\mathbb{Q}$. Thus, we conclude that Hensel's lemma is not applicable in $(\mathbb{Q}, ||)$.

Definition 6. A valued field $(K, ||)$ is said to be Henselian if Hensel's lemma is applicable in $(K, ||)$. The smallest Henselian field extending $(K, ||)$ is called the henselization of $(K, ||$ and denoted by K^h.

Exercices 5. Let $(K, ||)$ be a valued field $(K, ||)$.

Show that $K \subset K^h \subset \hat{K}$. Furthermore, these three fields have the same value group and same residue fields.

We have the following apparently easier characterization of Henselian fields. For the proof, we refer to ([16], Lemma 4.1.1).

Theorem 2.7. *The valued field* $(K, ||)$ *is Henselian if and only if it extends uniquely to* K^s, where K^s is the separable closure of K.

In particular, we conclude the following characterization of the henselization K^h of $(K, ||)$.

Theorem 2.8. *Let* $(K, ||)$ *be a valued field. Then* K^h *is the separable closure of* K *in* K^s.

3. Main results

Let $(K, ||)$ be a non archimidean valued field, ν the associated valuation to $||$ defined by $\nu(a) = -Ln|a|$ for every $a \in K^*$, $R_{||}$ its valuation ring, $M_{||}$ its maximal ideal, $\mathbb{F}_{||}$ its residue field, and (K^h, ν^h) its henselization.

3.1 Key polynomials

The notion of key polynomials was introduced in 1936, by MacLane [17], in the case of discrete rank one absolute values and developed in [18] by Vaquié to any arbitrary rank valuation. The motivation of introducing key polynomials was the problem of describing all extensions of $||$ to any finite simple extension $K(\alpha)$. For any simple algebraic extension of K, MacLane introduced the notions of key polynomials and augmented absolute with respect to the gievn key.

Definition 7. Two nonzero polynomials f and g in $R_{||}[x]$,

1. f and g are said to be $||$-equivalent if $|f - g|_\infty < |f|_\infty$.

2. We say that g is $||$-divides f if there exists $q \in R_{||}[x]$ such that f and gq are $||$-equivalent.

3. We say that a polynomial $\phi \in R_{||}[x]$ is $||$-irreducible if for every f and g in $R_{||}[x]$, ϕ $||$-divides fg implies that ϕ $||$-divides f or ϕ $||$-divides g.

Definition 8. A polynomial $\phi \in R_{||}[x]$ is said to be a MacLane-Vaquié key polynomial of $||$ if it satisfies the following three conditions:

1. ϕ is monic,

2. ϕ is $||$-irreducible,

3. ϕ is $||$-minimal; for every nonzero polynomial $f \in R_{||}[x]$, ϕ $||$-divides f implies that $\deg(\phi) \leq \deg(f)$.

It is easy to prove the following lemma:

Lemma 3.1. *Let $\phi \in R_{||}[x]$ be a monic polynomial. If $\overline{\phi}$ is irreducible over $\mathbb{F}_{||}$, then ϕ is a MacLane-Vaquié key polynomial of $||$.*

3.2 Augmented absolute values

Let $\phi \in R_{||}[x]$ be a MacLane-Vaquié key polynomial of $||$ and $\gamma \in \mathbb{R}^+$ with $\gamma \leq |\phi|_\infty$. Let $\omega : K(x) \to \mathbb{R}_{\geq 0}$, defined by $\omega(P) = \max\left\{|p_i|_\infty \gamma^i, i = 0, \dots, l\right\}$ for every $P \in K[x]$, with $P = \sum_{i=0}^{l} p_i \phi^i$ and $\deg(p_i) < \deg(\phi)$ for every $i = 0, \dots, l$ and extended by $\omega(A/B) = \omega(A) - \omega(B)$ for every nozero A and B of $K(x)$.

Lemma 3.2 *Let $P = \sum_{i=0}^{n} b_i \phi^i$ be a ϕ-expansion of P, where the condition $\deg(b_i) < \deg(\phi)$ for every $i = 0, \dots, n$ is omitted. If $\overline{\phi}$ does not divide $\overline{b_i(x)/b}$ for every $i = 0, \dots, n$, then $\omega(P) = \max\{|b_i|_\infty \gamma^i, i = 0, \dots, n\}$, where $b \in R_\nu$ such that $|b_i|_\infty = |b|$. Such an expansion is called an admissible expansion.*

Theorem 3.3. *Let $\phi \in R_{||}[x]$ be MacLane-Vaquié key polynomial of $||$ and $\gamma \in \mathbb{R}^+$ with $\gamma \leq |\phi|_\infty$. The map $\omega : K(x) \to \mathbb{R}_{\geq 0}$, defined by $\omega(P) = \max\left\{|p_i|_\infty \gamma^i, i = 0, \dots, l\right\}$*

for every $P \in K[x]$, with $P = \sum_{i=0}^{l} p_i \phi^i$ and $deg(p_i) < deg(\phi)$ for every $i = 0, \ldots, l$, and extended by $\omega(A/B) = \omega(A)/\omega(B)$ for every nonzero polynomials $(A, B) \in K[x]^2$, is an absolute value of $K(x)$.

Proof. It suffices to check that ω satisfies the three proprieties of an absolute value in $K[x]$. Let $(A, B) \in K[x]$ be tow polynomials, $A = \sum_{i=0}^{k} a_i \phi^i$, and $B = \sum_{i=0}^{s} b_i \phi^i$ the ϕ-expansions.

1. $\omega(A) = 0$ if and only if $|a_i|_\infty \gamma^i = 0$ for every $i = 0, \ldots, k$, which means $|a_i|_\infty = 0$ for every $i = 0, \ldots, k$ (because $\gamma \in \mathbb{R}^+$). Therefore $A = 0$.

2. Let $A = \sum_{i=0}^{l} a_i \phi^i$ and $B = \sum_{i=0}^{t} b_i \phi^i$ be the ϕ-expansions of A and B, with $\deg(a_i) < \deg(\phi)$ and $\deg(b_i) < \deg(\phi)$ for every $i = 0, \ldots, sup(l, t)$. For every $i = 0, \ldots, L = l + t$, let $c_i = \sum_{j=0}^{i} a_j b_{i-j}$. Then $AB = \sum_{i=0}^{L} c_i \phi^i$. For every $i = 0, \ldots, L$, upon the Euclidean division, let $c_i = q_i \phi + r_i$. Then $AB = \sum_{i=0}^{L} f_i \phi^i$ is the ϕ-expansion of AB, where $f_i = r_i + q_{i-1}$ with $q_{-1} = 0$. Since $|\phi|_\infty = 1$, we conclude that $|r_i|_\infty \leq |c_i|_\infty$ and $|q_i|_\infty \leq |c_i|_\infty$ for every $i = 0, \ldots, L$. Let i_1 and i_2 be the smallest integers satisfying $\omega(A) = |a_{i_1}|_\infty \gamma^{i_1}$ and $\omega(B) = |b_{i_2}|_\infty \gamma^{i_2}$. Then by the ultra-metric propriety $\omega(c_i \phi^i) \leq |a_{i_1}|_\infty \gamma^{i_1} |b_{i_2}|_\infty \gamma^{i_2}$ for every $i = 0, \ldots, L$. For the equality, by definition of i_1 and i_2, $\omega(a_{i_1} b_j \phi^{i_1 + j}) < |a_{i_1} b_j|_\infty \gamma^{i_1 + j}$ for every $j < i_2$ and $\omega(a_j b_{i_2} \phi^{j + i_2}) < |a_j b_{i_2}|_\infty \gamma^{j + i_2}$ for every $j < i_1$. Thus by using the expression of $c_{i_1 + i_2}$, we conclude the equality. For $i = i_1 + i_2$, let $(a, b) \in R_\nu^2$, with $|a_{i_1}|_\infty = |a|$ and $|b_{i_2}|_\infty = |b|$. If $|r_i|_\infty < |c_i|_\infty = |ab|$, then $|r_i/c|_\infty < |c_i/c|_\infty$, and so $|r_i/c|_\infty < 1$. By reducing modulo $M_{||}$, we deduce that $\overline{\phi}$ divides $\overline{c_i/c}$, which means that ϕ $||$-divides $(a_{i_1}/a)(b_{i_2}/b)$, which is impossible because ϕ is MacLane-Vaquié key polynomial of $||$, ϕ does not $||$-divide a_{i_1}/a, and ϕ does not $||$-divide $b_{i_2})/b$. Therefore, $|r_i|_\infty = |c_i|_\infty$. Hence $\omega(AB) = \max\{|a_i|_\infty \gamma^i, i = 0, \ldots, l\} \cdot \max\{|b_i|_\infty \gamma^i, i = 0, \ldots, t\}$ and $\omega(AB) = \omega(A)\omega(B)$ as desired.

3. Completing by zeros; $a_i = 0$ if $i > k$ and $b_i = 0$ if $i > s$, we have $\omega(A + B) = \max\{|a_i + b_i|_\infty \gamma^i, i = 0, \ldots, \sup(k, s)\} \leq \max\{|a_i|_\infty \gamma^i, i = 0, \ldots, k\} + \max\{|b_i|_\infty \gamma^i, i = 0, \ldots, s\} = \omega(A) + \omega(B)$. Thus, $\omega(A + B) \leq \omega(A) + \omega(B)$.

□

Definition 9. The absolute value ω defined in Theorem 3.3 is denoted by $[||\phi, \gamma,]$ and called the augmented absolue value of $||$ associated to ϕ and γ.

Example 3. Let $|||$ be the 2-adic absolute value defined on $\mathbb{Q}$ by $|a| = e^{-\nu_2(a)}$, where for every integer b, $\nu_2(b)$ is the largest integer satisfying 2^k divides b in $\mathbb{Z}$. Let $\phi = x^2 + x + 1 \in \mathbb{Z}[x]$. By Lemma 3.1, ϕ is a MacLane-Vaquié key polynomial of $||$. Since $|\phi|_\infty = 1$, for every real γ, $0 < \gamma \leq 1$, the map $\omega : \mathbb{Q}(x) \to \mathbb{R}_{\geq 0}$, defined by $\omega(P(\alpha)) = \max\{|p_i|_\infty \gamma^i, i = 0, \ldots, l\}$ for every $P \in K[x]$, with $P = \sum_{i=0}^{l} p_i \phi^i$ and $\deg(p_i) < 2$.

3.3 Extensions of absolute values

The following Lemma makes a one–one correspondence between the absolute value of L and monic irreducible factors of $f(x)$ in $K^h[x]$ for any simple finite extension $L = K(\alpha)$ of K generated by a root $\alpha \in \overline{K}$ of a monic irreducible polynomial $f(x) \in K[x]$.

Lemma 3.4. ([19], Theorem 2.1)

Let $L = K(\alpha)$ generated by a root $\alpha \in \overline{K}$ of a monic irreducible polynomial $f(x) \in K[x]$ and $f(x) = \prod_{i=1}^{t} f_i^{e_i}(x)$ be the factorization into powers of monic irreducible factors of $K^h[x]$. Then $e_i = 1$ for every $i = 1, \dots, t$ and there are exactly t distinct valuations $||_1, \dots,$ and $||_t$ of L extending $||$. Furthermore for every absolute value $||_i$ of L associated to the irreducible factor f_i, $|P(\alpha)|_i = \overline{|P(\alpha_i)|}$, where $\overline{||}$ is the unique absolute value of $\overline{K^h}$ extending $||$ and $\alpha_i \in \overline{K}$ is a root of $f_i(x)$.

Lemma 3.5. ([16], Corollary 3.1.4)

Let L/K be a finite extension and R_L the integral closure of $R_{||}$ in L. Then

$$R_L = \cap_{||_L} R_{||_L};$$

for any elemnt $\alpha \in L$, $\alpha \in R_L$ if and only if $|\alpha|_L \leq 1$ for every absolute value $||_L$ of L extending $||$.

Lemma 3.6. *Let $f(x) \in R_{||}[x]$ be a monic irreducible polynomial such that $\overline{f(x)}$ is a power of $\overline{\phi}$ in $\mathbb{F}_{||}[x]$ for some monic polynomial $\phi \in R_{||}[x]$, whose reduction is irreducible over $\mathbb{F}_{||}$. Let $L = K(\alpha)$ with $\alpha \in \overline{K}$ a root of $f(x)$. Then for every absolute value $||_L$ of L extending $||$, for every nonzero polynomial $P \in K[x]$, $|P(\alpha)|_L \leq |P|_\infty$.*

The equality holds if and only if $\overline{\phi}$ does not divide $\overline{P_0}$, where $P_0 = \frac{P}{a}$, with $a \in K$ such that $|P|_\infty = |a|$.

In particular, $|\phi(\alpha)|_L < 1$ and $|P(\alpha)|_L = |P|_\infty$ for every polynomial $P \in K[x]$ such $deg(P) < deg(\phi)$.

Proof. Let $||_L$ be an absolute value of L extending $||$, $P \in K[x]$ a nonzero polynomial, and $a \in K$ with $|a| = |P|_\infty$. Then $|P_0|_\infty = 1$. Since α is integral over $R_{||}$, we conclude that $|P_0(\alpha)|_L \leq 1$. Thus, $|P(\alpha)|_L \leq |a| = |P|_\infty$.

Moreover, the inequality $|P(\alpha)|_L < |P|_\infty$ means that $P_0(\alpha) \in M_{||_L}$, which means that $P_0(\alpha) \equiv 0 \pmod{M_{||_L}}$. Consider the ring homomorphism $\varphi : \mathbb{F}_{||}[x] \to \mathbb{F}_{M_{||_L}}$, defined by $\varphi(\overline{P}) = P(\alpha) + M_{||_L}$. Then $P_0(\alpha) \not\equiv 0 \pmod{M_{||_L}}$ is equivalent to $\overline{\phi}$ does not divide $\overline{P_0}$.

In particular, since $\phi \in R_{||}[x]$, $|\phi|_\infty \leq 1$. Furthermore as $\overline{\phi}$ divide $\overline{\phi}$, we conclude that $|\phi|_\infty < 1$.

Let $P \in K[x]$ be a nonzero polynomial of degree less than degree of ϕ. Then $P_0(x) \in R_{||}[x]$ is a primitive polynomial; $|P_0|_\infty = 1$. As degree $\overline{P_0}$ is less than degree of $\overline{\phi}$, $\overline{\phi}$ does not divide $\overline{P_0}$. Thus $|P(\alpha)|_L = |P|_\infty$. □

Theorem 3.7. *Let $f(x) \in R_{||}[x]$ be a monic polynomial. If $f(x)$ is irreducible over K^h, then $\overline{f(x)}$ is a power of $\overline{\phi}$ in $\mathbb{F}_{||}[x]$ for some monic polynomial $\phi \in R_{||}[x]$, whose reduction is irreducible over $\mathbb{F}_{||}$. Moreover if we set $f(x) = \sum_{i=0}^{n} a_i(x)\phi^i(x)$ the ϕ-expansion of $f(x)$, then $|a_{n-i}|_\infty \leq \gamma^i$ for every $i = 0, \dots, n$, where $\gamma = |a_0|_\infty^{1/n}$.*

Proof. The first point of the theorem is an immediate consequence of Theorem 2.6. For the second point, let $m = \deg(\phi)$.

1. For $m = 1$, let $f(x) = \prod_{i=1}^{k}(x - \alpha_i)$, where $\alpha_1, \dots, \alpha_k$ be the roots of $f(x)$ in $\overline{K}$, the algebraic closure of K. Then the formula linking roots and coefficients of $f(x)$, we conclude that $f(x) = \sum_{i=0}^{k} s_i x^i$, where $s_k = 1$, $s_i = \sum \prod_{j_1 < \dots < j_i} \alpha_{j_1} \cdots \alpha_{j_i}$. Keep the notation $||$ for the valuation of K^h extending $||$ and let $\overline{||}$ be the unique extension of $||$ to $\overline{K^h} = \overline{K}$. Then $\overline{|\alpha_1|} = \dots = \overline{|\alpha_k|} = \tau$, $\overline{|s_{k-i}|} \leq \tau^i$, and $\tau = \gamma$.

2. For $m \geq 2$, let $Ł = K^h(\alpha)$, where $\alpha \in \overline{K}$ is a root of $f(x)$, $g(x) = x^t + b_{t-1}x^{t-1} + \dots + b_0$ the minimal polynomial of $\phi(\alpha)$ over K^h, and $F(x) = g(\phi(x)) = \phi(x)^t + b_{t-1}\phi(x)^{t-1} + \dots + b_0$. By the previous case, we conclude that $|b_{t-i}| \leq \tau^i$ for every $i = 0, \dots, t$ with $\tau = |b_0|^{1/t}$, which means that $N_\phi(F) = S$ has a single side of slope $-\lambda = -\frac{\nu(b_0)}{t}$. Since $F(\alpha) = 0$, we conclude that $f(x)$ divides $F(x)$, and so $N_\phi(f)$ has a single side of the same slope $-\lambda$. Therefore, $|a_{n-i}|_\infty \leq \gamma^i$ for every $i = 0, \dots, n$, where $\gamma = |a_0|_\infty^{1/n}$.

□

Exercices 6. Let $(K, |\,|)$ be a non archimidean valued field and $f(x) \in K^h[x]$. Set $f(x) = \sum_{i=0}^n a_i(x)\phi^i(x)$ the ϕ-expansion of $f(x)$.

Show that $|f(x)|_\infty = \max(|a_n|_\infty, |a_0|_\infty)$.

Based on absolute value, the following theorem gives an hyper bound of the number of monic irreducible factors of monic polynomials. In particular, Corollary 3.9 gives a criterion to test the irreducibility of monic polynomials.

Theorem 3.8. *Let $(K, |\,|)$ be a non archimidean valued field, $\Gamma = |K^*|$ its value group, and $f(x) \in K[x]$ a monic polynomial such that $\overline{f(x)}$ is a power of $\overline{\phi}$ in $\mathbb{F}_{|\,|}[x]$. Let $f(x) = \sum_{i=0}^n a_i(x)\phi^i(x)$ be the ϕ-expansion of $f(x)$ and assume that $|a_{n-i}|_\infty \leq \gamma^i$ for every $i = 0, \dots, n$, where $\gamma = |a_0|_\infty^{1/n}$. Let e be the smallest positive integer satisfying $\gamma^e \in \Gamma$. Then $f(x)$ has at most d irreducible monic factors in $K^h[x]$, where $d = n/e$ with degree at least em each, and $m = deg(\phi)$.*

Proof. By applying the map $-Ln$, the hypothesis $|a_{n-i}|_\infty \leq \gamma^i$ for every $i = 0, \dots, n$ means that $\nu(a_{n-i}) \geq i\lambda$, where $\lambda = \frac{\nu(a_0)}{n}$, which means that $N_\phi(f) = S$ has a single side of slope $-\lambda$ with respect to ν. Let $f(x) = \prod_{i=1}^t f_i(x)$ be a non trivial factorization of monic polynomials in $K^h[x]$. Then by Theorem 2.2, $N_\phi(f_i) = S_i$ has a single side of slope $-\lambda$. Fix $i = 1, \dots, t$ and let $f_i(x) = \sum_{j=0}^{l_i} a_{ij}(x)\phi^j$ be the ϕ-expansion of f_i. Then $\deg(f_i) = l_i m$ and $-Ln(\gamma) = -\lambda$ is the slope of S_i. Since e is the smallest positive integer satisfying $\gamma^e \in \Gamma$, we conclude that e is the smallest positive integer satisfying $e\lambda \in \nu(K^*)$. On the other hand, since $\lambda = \frac{a_{i0}}{l_i}$ is the slope of S_i, where l_i is the length of the side S_i, we conclude that e divides l_i. Thus $\deg(f_i) = d_i em$, where $d_i = \frac{l_i}{e}$. It follows that every non trivial factor $f_i(x)$ has degree at least em. Since $\deg(f) = \sum_{i=1}^t \deg(f_i) \geq tem$, we conclude that $t \leq \frac{n}{e} = d$. □

Corollary 3.9. *Under the hypothesis and notations of Theorem 3.8, if $e = n$, then $f(x)$ is irreducible over K^h.*

Proof. If $n = e$, then $d = 1$, and so there is a unique monique polynomial of $K[x]$ which divides $f(x)$ and this factor has the degree at least mn. As $\deg(f) = nm$, we conclude that $f(x)$ is this unique monic factor. □

Theorem 3.10. *Let $L = K(\alpha)$ be a simple extension generated by $\alpha \in \overline{K}$ a root of a monic irreducible polynomial $f(x) \in R_{|\,|}[x]$ such that $\overline{f(x)} = \overline{\phi}^n$ in $\mathbb{F}_{|\,|}[x]$. Let $f(x) = \sum_{i=0}^n a_i(x)\phi^i(x)$ be the ϕ-expansion of $f(x)$. Assume that $|a_{n-i}|_\infty \leq \gamma^i$ for every $i = 0, \dots, n$, where $\gamma = |a_0|_\infty^{1/n}$. Then for every absolute value $|\,|_L$ of L extending $|\,|$, $|P(\alpha)|_L \leq max\left\{|p_i|_\infty \gamma^i, i = 0, \dots, l\right\}$ for every $P \in K[x]$, with $P = \sum_{i=0}^l p_i\phi^i$ and $l < n$.*

Proof. Let $|\,|_L$ be an absolute value of L extending $|\,|$ and let us show that $|\phi(\alpha)|_L = \gamma$. For this reason, let $\tau = |\phi(\alpha)|_L$. By Lemma 3.6, $0 < \tau < 1$. By hypotheses and Lemma 3.6, $\left|a_i(\alpha)\phi(\alpha)^i\right|_L \leq \gamma^{n-i}\tau^i$ for every $i = 0, \dots, n$. Thus, if $\tau \neq \gamma$, $\max\left\{\left|a_i(\alpha)\phi(\alpha)^i\right|_L, i = 0, \dots, n\right\} = \max(\tau^n, \gamma^n)$. Since $||$ is a non archimidean absolute value, we conclude that $||_L$ is a non archimidean absolute value, and so by the ultra-metric propriety, $|f(\alpha)|_L = \max(\tau^n, \gamma^n) > 0$, which is impossible because $f(\alpha) = 0$. Therefore $|\phi(\alpha)|_L = \gamma$.

Now, let $P = \sum_{i=0}^{l} p_i\phi^i$ be a polynomial in $K[x]$. By the ultra-metric propriety, $|P(\alpha)|_L \leq |p_i(\alpha)|_L \gamma^i$. □

Theorem 3.11. *Let $L = K(\alpha)$ be a simple extension generated by $\alpha \in \overline{K}$ a root of a monic irreducible polynomial $f(x) \in R_{||}[x]$ such that $\overline{f(x)} = \overline{\phi}^n$ in $\mathbb{F}_{||}[x]$. Let $f(x) = \sum_{i=0}^{n} a_i(x)\phi^i(x)$ be the ϕ-expansion of $f(x)$. Assume that $|a_{n-i}|_\infty \leq \gamma^i$ for every $i = 0, \dots, n$, where $\gamma = |a_0|_\infty^{1/n}$. If n is the smallest positive integer satisfying $\gamma^e \in \Gamma$, then there is a unique absolute value $||_L$ of L extending $||$. Moreover this absolute value is defined by $|P(\alpha)|_L = \max\left\{|p_i|_\infty \gamma^i, i = 0, \dots, l\right\}$ for every $P \in K[x]$, with $P = \sum_{i=0}^{l} p_i\phi^i$ and $l < n$.*

Furthermore, its ramification index is n and its residue degree is $m = \deg(\phi)$.

Proof. By Corollary 3.9, if $n = e$, then $f(x)$ is irreducible over K^h. Thus by Hensel's Lemma, there is a unique absolute value $||_L$ of L extending $||$. By Theorem 3.10, we conclude that $|\phi(\alpha)|_L = \gamma$ and $|P(\alpha)|_L \leq |p_i(\alpha)|_L \gamma^i = |p_i|_\infty$ for every polynomial $P = \sum_{i=0}^{l} p_i\phi^i$ in $K[x]$. Let us show the equality. Let s be the smallest integer which satisfies $\omega(P) = |p_s|_\infty \gamma^s$. Let i be an integer satisfying $\omega(P) = |p_i|_\infty \gamma^i$. Then $\gamma^{s-i} = |p_i|_\infty / |p_s|_\infty \in \Gamma$. Thus n divides $i - s$ because n is the smallest positive integer satisfying $\gamma^e \in \Gamma$. Since $l < n$, then $i - s = 0$. Therefore, $|P(\alpha)|_L = |p_s|_\infty \gamma^s = \omega(P)$.

For the residue degree and ramification index, since $|\phi(\alpha)|_L = \gamma$ and $n = (\Gamma(\gamma) : \Gamma)$, we conclude that n divides the ramification index e of $||_L$. On the other hand, since $\mathbb{F}_{||} \subset \mathbb{F}_\phi \subset \mathbb{F}_{||_L}$, with $\mathbb{F}_\phi = \frac{\mathbb{F}_{||}[x]}{(\overline{\phi})}$, we have $m = [\mathbb{F}_\phi : \mathbb{F}_{||}]$ divides $[\mathbb{F}_{||_L} : \mathbb{F}_{||}]$. As $m \cdot n = \deg(f)$, we conclude the equality. □

Exercices 7. For every positive integer $n \geq 2$ and p a positive prime integer, let $f(x) = x^n - p$.

1. How that f is irreducible over $\mathbb{Q}$.

2. Conclude a new proof for $[\mathbb{R} : \mathbb{Q}] = \infty$.

3. Let $L = \mathbb{Q}(\alpha)$ with α a complex root of $f(x)$. Show that there is a unique absolute value of L extending the absolute $||_p$, defined on $\mathbb{Q}$ by $|a|_p = e^{-\nu_p(a)}$ for every $a \in \mathbb{Q}$, where ν_p is the p-adic valuation on $\mathbb{Q}$. Calculate its residue degree and its ramification index.

Combining Lemma 3.4 and Theorem 3.8, we conclude the following result:

Corollary 3.12. *Let $L = K(\alpha)$ be a simple extension generated by $\alpha \in \overline{K}$ a root of a monic irreducible polynomial $f(x) \in R_{||}[x]$. Let $\overline{f(x)} = \prod_{i=1}^{r} \overline{\phi_i}^{n_i}(x)$ be the factorization of $\overline{f(x)}$ in $\mathbb{F}_{||}[x]$, with every $\phi_i \in R_{||}[x]$ is a monic polynomial. For every $i = 1, \dots, r$, let $N_{\phi_i}^+(f) = S_{i1} + \dots + S_{ig_i}$ be the principal ϕ_i-Newton polygon of $f(x)$. Then L has t absolue value extending $||$ with $r \leq t \leq \sum_{i=1}^{r}\sum_{j=1}^{g_i} d_{ij}$, where $d_{ij} = \frac{l_{ij}}{e_{ij}}$ is the degree of S_{ij}, l_{ij} is the length of S_{ij}, and $e_{ij} = \frac{l_{ij}}{d_{ij}}$ for every $i = 1, \dots, r$ and $j = 1, \dots, g_i$.*

4. Applications

1. Let || be the p-adic absolute value defined on $\mathbb{Q}$ by $|a| = p^{-\nu_p(a)}$ and $f(x) = x^n - p\mathbb{Z}[x]$. Show that $f(x)$ is irreducible over $\mathbb{Q}$. Let $L = \mathbb{Q}(\alpha)$ with α a complex root of $f(x)$. Determine all absolute value of L extending $||$.

   Answer. First $\Gamma = \{p^k, k \in \mathbb{Z}\}$ is the value group of $||$. Since $|p| = p^{-1}$, $\gamma = |a_0|^{1/n} = p^{-1/n}$, we conclude that the smallest integer satisfying $\gamma^e \in \Gamma$ is n. Thus, by Corollary 3.9, $f(x)$ is irreducible over $\mathbb{Q}^h$,and so is over $\mathbb{Q}$. Since $\overline{f(x)} = x^n$ in $\mathbb{F}_p[x]$, by Theorem 3.11, there is a unique absolute value of L extending $||$ and it is defined by $|P(\alpha)|_L = \max\{|p_i|\gamma^i, i = o, \dots, l\}$ for every polynomial $P = \sum_{i=0}^{l} x^i$ with $l < n$.

2. Let || be the p-adic absolute value and $f(x) = x^{n^{-a}} \in \mathbb{Z}[x]$ such that p does not divide $\nu_p(a)$. Show that $f(x)$ is irreducible over $\mathbb{Q}$. Let $L = \mathbb{Q}(\alpha)$ with α a complex root of $f(x)$. Determine all absolute value of L extending $||$.

   Answer. First $\Gamma = \{p^k, k \in \mathbb{Z}\}$ is the value group of $||$. Since $|p| = p^{-1}$, $\gamma = |a_0|^{1/n} = p^{-1/n}$, we conclude that the smallest integer satisfying $\gamma^e \in \Gamma$ is n. Thus, by Corollary 3.9, $f(x)$ is irreducible over $\mathbb{Q}^h$,and so is over $\mathbb{Q}$. Since $\overline{f(x)} = x^n$ in $\mathbb{F}_p[x]$, by Theorem 3.11, there is a unique absolute value of L extending $||$ and it is defined by $|P(\alpha)|_L = \max\{|p_i|\gamma^i, i = o, \dots, l\}$ for every polynomial $P = \sum_{i=0}^{l} x^i$ with $l < n$.

3. Let $f(x) = \phi^6 + 24x\phi^4 + 24\phi^3 + 15(16x + 32)\phi + 48$ with $\phi \in \mathbb{Z}[x]$ a monic polynomial whose reduction is irreducible in $\mathbb{F}_2[x]$. In $\mathbb{Q}_2[x]$, how many monic irreducible factors $f(x)$ gets?, where $\mathbb{Q}_2$ is the completion of $(\mathbb{Q}, ||)$ and $||$ is the 2-adic absolute value.

   Answer. It is easy to check that $f(x)$ satisfies the conditions of Theorem 3.8; $|a_{6-i}|_\infty \le \gamma^i$ with $\gamma = 2^{-4/6} = \left(2^{-1/3}\right)^2$. Thus $e = 3$ and $d = 2$. By Theorem 3.8, $f(x)$ has at most 2 monic irreducible factors in $\mathbb{Q}_2[x]$.

Author details

Lhoussain El Fadil* and Mohamed Faris
Faculty of Sciences Dhar El Mahraz, Sidi Mohamed ben Abdellah University, Morocco

*Address all correspondence to: lhouelfadil2@gmail.com

References

[1] G. Eisenstein, *Über die Irreduztibilität und einige andere Eigenschaften der Gleichungen, von welcher die Theilung der ganzen Lemniscate abhängt*, J. reine angew. Math. 39 (1850) 160–179.

[2] K. Girstmair, *On an irreducibility criterion of M. Ram Murty*, Amer. Math. Monthly 112 (2005) 269–270

[3] S. K. Khanduja and M. Kumar, *Prolongations of valuations to finite extensions*, Manus. Math. **131** (2010) 323–334.

[4] M. R. Murty, *Prime numbers and irreducible polynomials*, Amer. Math. Monthly 109 (2002) 452–458.

[5] S. H. Weintraub, *A family of tests for irreducibility of polynomials*, Proc. Amer. Math. Soc. 144 (2016) 3331–3332.

[6] K. Hensel, *Untersuchung der Fundamentalgleichung einer Gattung reine reelle Primzahl als Modul und Bestimmung der Theiler ihrer Discriminante*, J. Reine Angew. Math. 113 (1894) 61–83.

[7] O. Ore, *Newtonsche Polygone in der Theorie der algebraischen Körper*, Math. Ann., 99 (1928), 84–117

[8] G. Dumas, *'Sur quelques cas d'irréductibilité des polynômes à coefficients rationnels'*, J. Math. Pures Appl. 6 (1906) 191–258.

[9] R. Brown, *Roots of generalized Schönemann polynomials in henselian extension fields*, Indian J. Pure Appl. Math. 39(5) (2008) 403–410

[10] A. Deajim and L. El Fadil, *On the extensions of discrete valuations in number fields*, Mathematica Slovaca 69(5) (2019), 1009–1022

[11] L. El Fadil, *On prolongations of rank one discrete valuations*, Comment. Math. Univ. Carolin. 60,3 (2019), 299–304

[12] J. Neukirch, *Algebraic Number Theory*, Springer-Verlag, Berlin, 1999.

[13] J. Guardia, J. Montes, and E. Nart, *Newton polygons of higher order in algebraic number theory*, Trans. Amer. Math. Soc. **364** (1) (2012) 361–416.

[14] D. Cohen, A. Movahhedi and A. Salinier, *Factorization over local fields and the irreducibility of generalized difference polynomials*, Mathematika, **47** (2000), 173–196

[15] L. El Fadil, *On Newton polygon's techniques and factorization of polynomial over henselian valued fields*, J. of Algebra and its Appl. 19(10) (2020) 2050188

[16] J. Engler and A. Prestel, Valued Fields, Springer Berlin Heidelberg New York, 2005.

[17] *S. MacLane*: A construction for absolute values in polynomial rings. Trans. Amer. Math. Soc. **40** (1936), 363–395

[18] M. Vaquié, Extension d'une valuation, Trans. Amer. Math. Soc. 359 (7) (2007), 3439–3481

[19] A. Deajim, L. El Fadil, and A. Najim, *On a theorem of Dedekind* (arxiv preprint and to appear in Siberian Math. Journal)

Chapter 7

The Efficiency of Polynomial Regression Algorithms and Pearson Correlation (r) in Visualizing and Forecasting Weather Change Scenarios

Okba Weslati, Samir Bouaziz and Mohamed Moncef Serbaji

Abstract

In this chapter, we will discuss the application of Python using the polynomial regression approach for weather forecasting. We will also evoke the role of Pearson correlation in modifying the trend of climate forecast. The weather data were processed via Aqua Crop by introducing daily climate observations. Accordingly, the software outputs are: reference evapotranspiration, maximum and minimum temperature, and precipitation. Additionally, we focused on the interference of the input data on the efficiency of predicting climate change scenarios. For that matter, we used this machine learning algorithm for two case studies, depending on the type of input data. As a result, we found that the outcome of polynomial regression was very sensitive to those input factors.

Keywords: python, polynomial regression, Pearson correlation, AquaCrop, weather forecasting

1. Introduction

Weather forecast is playing a vital role in society, environment, and sociable development. Their utilities have exceeded the simple mission of providing information for the users about the weather behavior during the upcoming period. Indeed, the whole concept was changing where it has been applied recently for politics and deciders to reduce socioeconomic losses that could be potentially generated by climate, which plays a substantial role in assuring life quality and economic prosperity. For example, the United States has cited 96 natural disasters that occurred between 1980 and 2006, with total losses that exceed 700 billion dollars. Up to now, around 629 fatalities per year are directly caused by weather disasters. Moreover, more than 60,000 premature deaths are recorded annually and are originally caused by poor air quality. Additionally, more than 1.5 million road crashes are originally caused by weather causing 7400 deaths, 700,000 injuries, and around 42 billion losses. Economically, more than 42 billion dollars are estimated to be lost due to weather traffic delays [1]. On the other hand, researchers are using weather forecasting to set new strategies that benefit the most from weather changes. Environmentalists are rather focusing on protecting the ecosystem from any possible climate change threats.

The services of weather are rising sharply. Gained profits from weather forecasts and warning disasters have reached over 31 billion dollars. Eventually, good forecasting will automatically lead to earlier warming and thus to good precautions, which will eventually contribute to reduce weather fatalities and economic loss. Technics of observing and studying weather behavior are continuously progressing, but understanding how the climate reacted in the future is still often too hard to simulate. Related to this point, the diversity of machine learning algorithms has led to predict various climate responses, providing more and more accurate forecasts. Since then, details that deal with the numerical and spatial resolution are being continuously developing, acquiring more challenging computing capabilities.

In the meanwhile, the climate is still too sensitive and chaotic. Even so, it is often too hard to obtain the perfect forecast, but ideally, every weather forecast model needs to get a certain measure of confidence, depending on many external parameters (type of data, coordinates, altitude, etc.). Generally, the best forecast model is the one that implements voluminous data and various methods/models. We must therefore limit the potential parameters that influence mostly the behavior of climate. As a consequence, machine learning algorithms can generate multiple forecast scenarios with slightly different initial inputs and/or changes in some stochastic parametrizations or model formulations. The forecast uncertainty is dependent mostly on the initial states of the atmosphere where it was observed, plus a certain random factor. Any changes in the format of weather data can modify the real observed status of the atmosphere at a certain point. Consequentially, a generated model needs much time between processing and delivering to the users.

Machine learning is a field of computer science based on artificial intelligence. The whole concept is based on provide learning capability to the users (computers or other devices) without being explicitly operated. It aims to conceive a suitable models and algorithms to learn and forecast based on input data [2, 3]. In addition, machine learning algorithms are efficiently used to describe the behavior of the dataset, dealing with noisy and nonstationary data. It uses a model input features by producing an expected output and forecasts suitable output features established on its historical records. Given the wide availability of weather data, fast and accurate decision-making is becoming a vital and more important than ever. Therefore, machine learning algorithms are one of the best alternatives of forecasting weather behavior. Besides, they can easily adapt themselves to changing trends inside datasets and can thus generate models based on input data instead of applying a conventional generalized model.

Many research studies have spoken about weather forecasting using several analysis methods. Preliminary studies about weather forecasting were based on persistence and statistical methods such as regression models [4]. These models are the most common approaches that use statistical technique for weather prediction. One of the most famous regression models is polynomial approach, which provides an effective way to describe complex and voluminous dataset in a nonlinear form. Additionally, polynomial regression models are based on the observed relationship between the dependent and independent variables to find out the most suitable polynomial equation order.

Therefore, the present chapter lays out the outcome of using the polynomial regression models for weather forecasting and to expose all the factors and/or parameters that can potentially affect the efficiency of the prediction. We will reveal the risk of dealing with big data volumes and how the format of the input data will affect the accuracy of the model. We will review the application of the Pearson correlation in retrieving or homogenizing the data and how it can affect the prediction climate behavior. We will discuss the best polynomial algorithm that fits mostly with the type of the data and its ability to generate a valid, concrete and unquestionable weather forecast model.

2. Material and data

Meteorological data are derived originally from two sources. In the first case, daily climate data were collected and processed in AquaCrop to calculate a yearly average for every one of the following parameters: maximum and minimum temperature (T_{max} and T_{min}), reference evapotranspiration (ET_0), and precipitation (P). These data were assembled for the watershed of Mellegue, between 2002 and 2019. In the second case, we collected monthly precipitation for the city of Zaghouan (Tunisia) from CHIRPS website for the corresponding period of 1981–2019. We want to test the efficiency of machine learning algorithms toward the type of input data. Later, precipitation data were used into the polynomial regression equation to predict the rain conduct for the next decade (2020–2030).

2.1 Polynomial regression

2.1.1 Method

The main algorithm used in this study is polynomial regression. It has been widely applied, and its statistical tools are famous [5]. Generally speaking, it is a form of linear regression that is why we do call it sometimes "polynomial linear regression model." It is a form of regression analysis that links the independent variable (x) to the dependent variable (y) according to the nth polynomial degree [4]. The general equation of polynomial regression is written in the following form:

$$y = \beta_0 + \beta_1 x + \beta_2 x^2 + \ldots \beta_n x^n \tag{1}$$

β_1 = linear effect parameter.

β_2 = quadratic effect parameter.

β_0 = a constant parameter, which is determined according to the polynomial function when $x = 0$.

Hence, we can find in some polynomial equations a certain factor (θ) called residual error.

In another way, we can define polynomial regression as a form of linear regression between dependent and independent variables where we add some terms or factors (based on a curvilinear relationship) to convert it into a polynomial regression. Nonetheless, we can simply return to the linear regression model. In that case, the previous equation will be:

$$y = \beta_0 + \beta_1 x \tag{2}$$

2.1.2 Why polynomial regression?

The general behavior of any climate data is set to be a nonlinear way. As a result, the linear regression model will be too difficult to visualize or predict entirely the data. For that reason, it will be too hard to draw the best line that fits mostly the weather data. The performance of the model will be too far from reality, and the projection of the

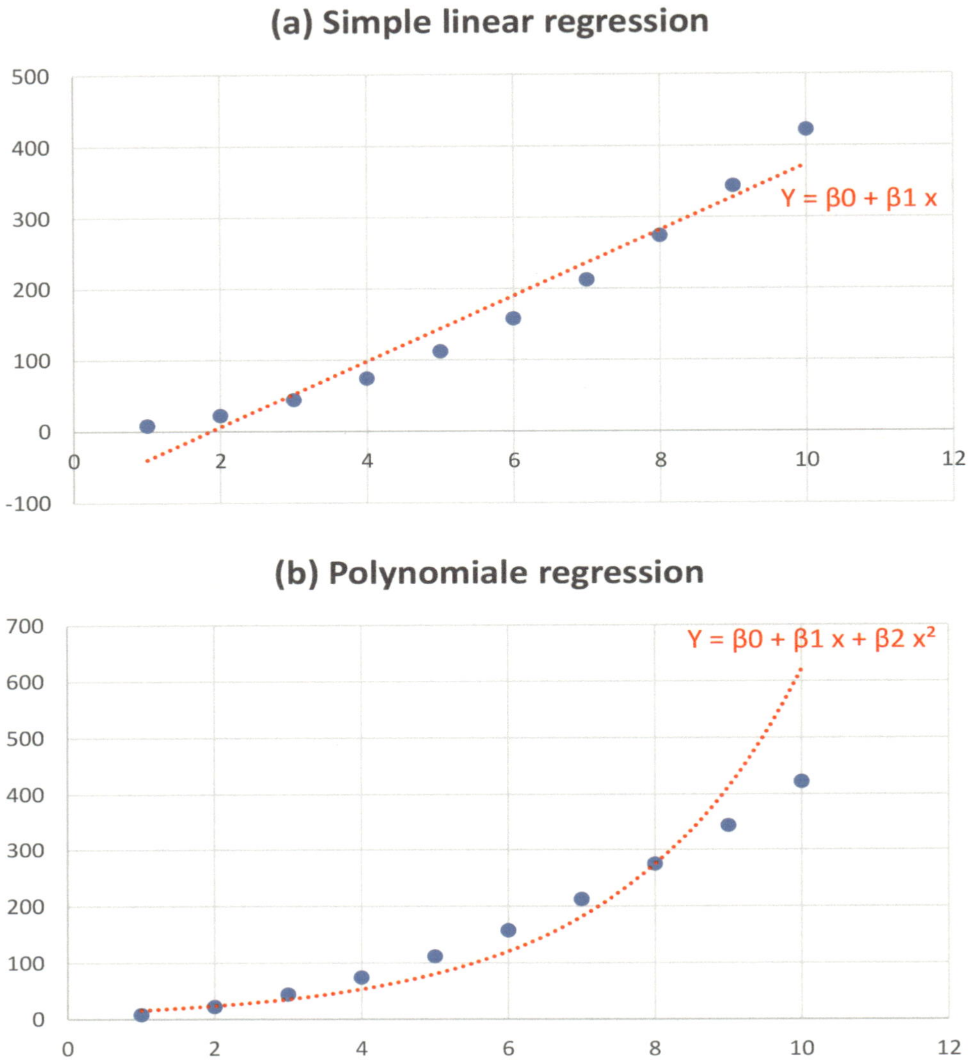

Figure 1.
Difference between linear and polynomial regression.

weather forecast will be, consequently, too doubtful. In this case, we opt for the polynomial regression to fit the data graph with a low value of error (**Figure 1**).

2.1.3 How to build polynomial regression in Python?

Python has several methods for finding the curvilinear relationship between data points. Based mainly on mathematic equations, the algorithm will succeed in drawing the best polynomial regression line that fits the most with the original input data. Later on, we can use the collected information to predict future values with a specific timescale. In Python, many packages were imported to operate this algorithm, but the main needed libraries are: "pandas," "matplotlib," "numpy," and "sklearn."

The process is quite easy with Python (**Figure 2**). The first step consists of importing major libraries especially "Pandas," "numpy," and "sklearn." You need to import the CSV file, which must have at least two columns; the column A contains the date (you have the right to put the conventional format of date you want, but be aware that it must be understood by Python!), and the second column that contains

DOI: http://dx.doi.org/10.5772/intechopen.102726

```
# -*- coding: utf-8 -*-
"""
Created on Wed Sep 29 15:51:54 2021

@author: HP
"""


#Polynomial Regression
from sklearn.preprocessing import PolynomialFeatures
df = pandas.read_csv("C:/Users/HP/Documents/private/Python_Zaghouan/zaghouan_rainfall_7.csv" , header=[0] )
x = np.array(df['Month']).reshape (-1,1)
print (x)
y = np.array(df['Rainfall'])
print (y)
x_ = PolynomialFeatures(degree=2, include_bias=False).fit_transform(x)
model = LinearRegression().fit(x_, y)
print(x_)
r_sq = model.score(x_, y)
# statistic of this quadratic polynomial regression
print('coefficient of determination:', r_sq)
print('intercept:', model.intercept_)
print('coefficients:', model.coef_)
y_pred = model.predict(x_)
print('predicted response:', y_pred, sep='\n')
plt.plot(df['Month'], df['Rainfall'])
plt.plot(x, y_pred)
plt.show

#forecasting weather according to the number of months
x_new= np.arange(468,589).reshape((-1,1))
y_new= model.predict(PolynomialFeatures(degree=4, include_bias=False).fit_transform(x_new))
print (y_new)
plt.plot(x_new, y_new)
```

Figure 2.
Script of polynomial regression model in python.

the observed values (which corresponds to the climate stats in our cases). Next, you have to specify the corresponding column to every variable and order the value gradually, by using the imported library "numpy." Going this far, all you have to do next is to apply the corresponded polynomial order that ideally fits with the observed data. You can assess the behavior of the polynomial regression graphically (using the libraries matplotlib) or statically (R^2 and "model. summary"). If you find the perfect model order, you can go further by forecasting the results at a specific time period.

2.1.4 How to judge the efficiency of polynomial regression?

- Statically: based on R^2 (coefficient of determination)

By definition, R^2 is an indicator that measures the quality of the prediction by linear regression. It is also known as the Pearson linear determination coefficient. It measures the fitness accuracy between the model and the observed data or how well the regression equation is suited to describe the distribution of sampling points. The coefficient varies from 0 to 1, referring to the strength of the prediction model. Meaning if R^2 is zero, the chosen mathematical model (based on the linear regression) does not succeed in describing and/or fitting the distribution of the points. In the opposite, if R^2 is 1 (or close to 1), we can conclude that selected regression line is able to determine the entire (100%) point cloud and the ability of the chosen mathematic equation in describing the points distribution. To conclude with, the closer R^2 is to zero, the more scatter plot will be dispersed around the regression line. Inversely, if R^2 tends to 1, it indicates that cloud points will congregate/narrow around the regression line.

- Graphically: based on back-forecasting

When we say predict, it is often related to anticipate the future response based on past or actual values. Instead, we want to compare the forecast equation with real data. So, here we do invent the conception of back-forecasting. Generally, any time series model looks the same by going forward or backward and can, therefore, predict the future as well as the past. Meaning instead of going forward, we use the model to go backward in time and predict the past. The model uses two backwards passes. The first one is used to calculate early data by the estimating parameters while the second one is made to establish the forecasting equation to calculate the forward values (future). In **Figure 3**, we used the back-forecasting method to squeeze most information out of the 468 monthly data (1981 to 2019). The more back-forecasted graph is close to the real data plot, the more accurate the forecasting equation will be.

2.2 Pearson correlation

Also known as the bivariate correlation, it is used to measure the strength of two sets of data based on the linear relationship. Mathematically speaking, it refers to the ratio between the covariance and the standard deviation of two variables (Eq. (3)). As such formula, the resultant value will always set between −1 and 1 where values close to 1 are referring to good correlation as the non-correlated values are close to zero. Generally speaking, the absolute values of the given correlation will be automatically linked to the strength of the values (negative values that are close to −1 are indicating that the two variables are inversely correlated, meaning that the rise of the first variable will lead to a decrease of the other).

The general equation of r ($\rho_{x,y}$) is written in the following formula:

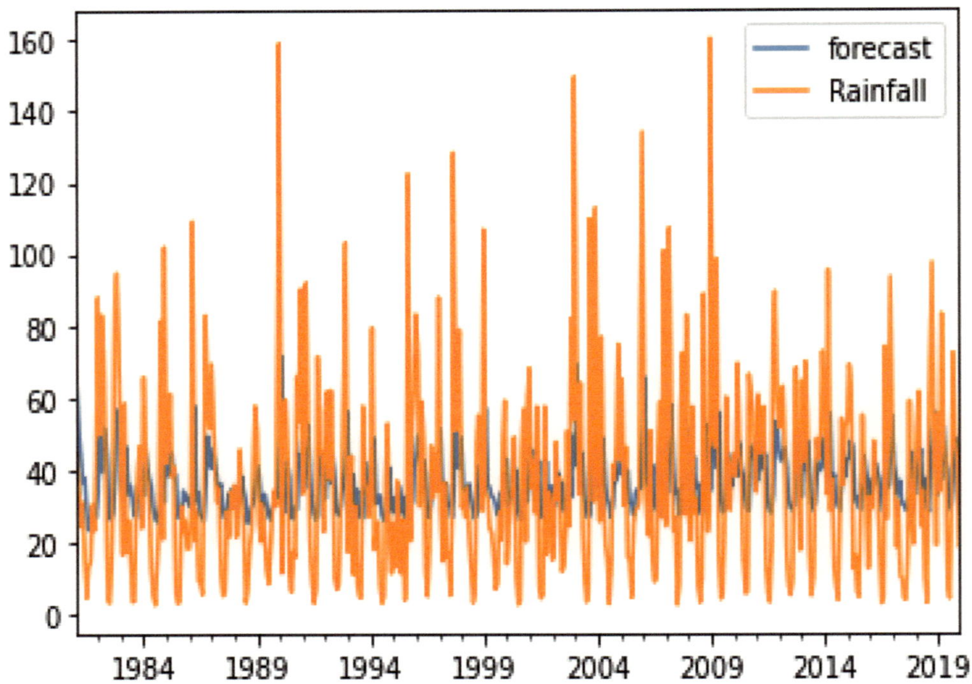

Figure 3.
Figure assimilation between back-forecasting plot (based on ARIMA (1,1,1)) and real rainfall values of Zaghouan.

$$\rho_{x,y} = \frac{\text{cov}\,(X,Y)}{\sigma_x \sigma_y} \tag{3}$$

Where $\rho_{x,y}$: r correlation.
Cov(x,y): covariance.
σ: Standard deviation.

We have to test the relationship between the data based on the Pearson coefficient. By proceeding this correlation, we have neglected all data from the stations that have a correlation less than 0.6 as it is recommended in some studies [6]. Lately, we have to rerun the polynomial regression for the validated data that has only a strong correlation. The program is too simple with Python, you can find here (**Figure 4**) the script of the operating Pearson correlation and how to export the results into a CSV file.

2.3 Data type

2.3.1 Yearly data: Mellegue catchment

The watershed of Mellegue is situated between 36° 25′ 50.43″, 35° 12′ 20.74″ north and 7° 11′ 30.98″, 8° 55′ 7.99″ east (**Figure 5**). The area covers 10,500 km^2 where approximately 60% of the surface belongs to Algeria, whereas the remaining surface, as well as the outlet, is situated in Tunisia. The local climate is known as arid to semiarid with a slight change in sub-humid in the north. As a consequence, the area is covered with low vegetation where few forests are situated mainly in the northeast of the basin. The average yearly temperature is around 17°c with a slight change of 1–2°C. The winter is responsible for providing 50% of the annual rainfall. The so-called Oued Mellegue represents the main river of the catchment. It is 290 km long, cuts the watershed in a northeastern southwestern direction. Due to its large area, we assembled weather data from 23 stations scattered all around the

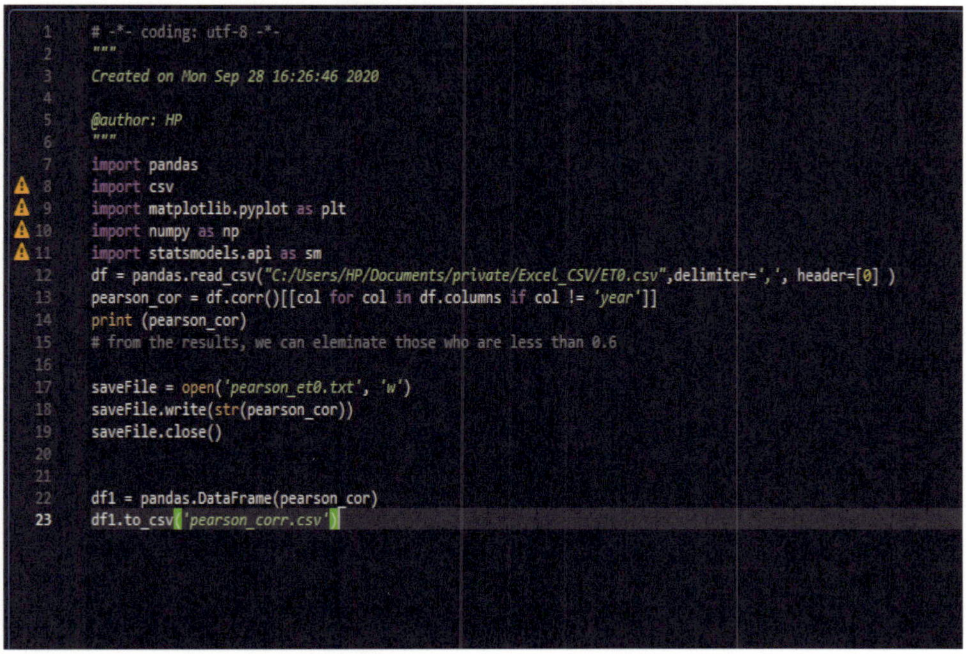

```
# -*- coding: utf-8 -*-
"""
Created on Mon Sep 28 16:26:46 2020

@author: HP
"""
import pandas
import csv
import matplotlib.pyplot as plt
import numpy as np
import statsmodels.api as sm
df = pandas.read_csv("C:/Users/HP/Documents/private/Excel_CSV/ET0.csv",delimiter=',', header=[0] )
pearson_cor = df.corr()[[col for col in df.columns if col != 'year']]
print (pearson_cor)
# from the results, we can eleminate those who are less than 0.6

saveFile = open('pearson_et0.txt', 'w')
saveFile.write(str(pearson_cor))
saveFile.close()


df1 = pandas.DataFrame(pearson_cor)
df1.to_csv('pearson_corr.csv')
```

Figure 4.
Script of the application of Pearson correlation in python.

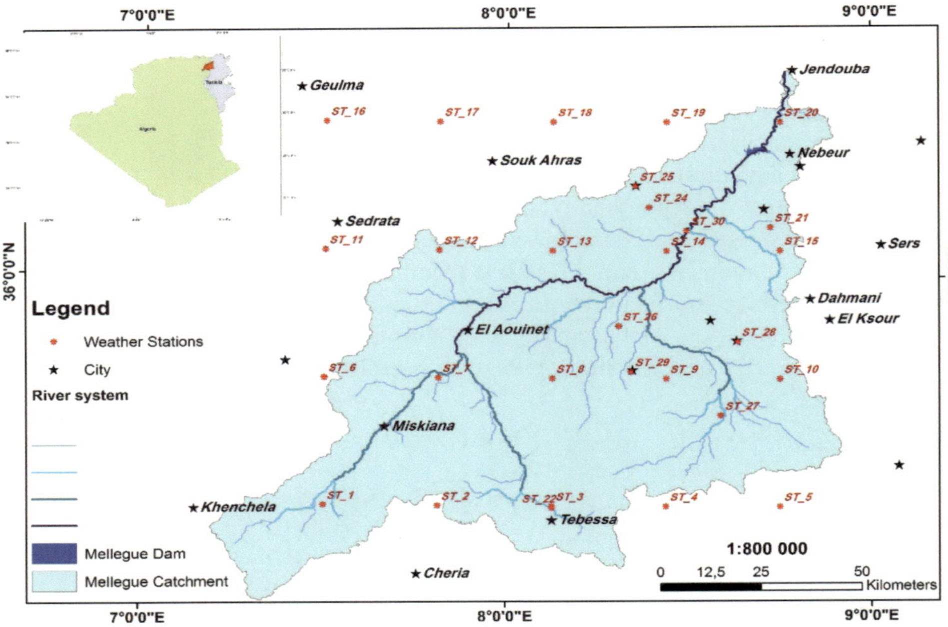

Figure 5.
Geographic location of the watershed of Mellegue.

catchment. Later on, the large amount of data was combined to compute a yearly average value for each weather variable. By homogenizing all the weather data together, it will be very helpful in understanding and visualizing the response of the watershed due to climate change and relieving voluminous data processing by the model. Hence, dealing with each station separately will lead to a lack in predicting the climate behavior of the area [7].

2.3.2 Monthly data: the city of Zaghouan

The city of Zaghouan is a Tunisian metropolitan located under the coordinates 36° 24′ north/10° 09′ east (**Figure 6**). The city is situated in the northeast of Tunisia, specifically on the hill of the Jbel Zaghouan (1295 m altitude). The region is known for the abundance of high reliefs and a large number of water sources because of the active seismicity of the zone, mainly manifested in the fault of Zaghouan. This mechanic structure extends for approximately 80 km along with the northeast-southwest Atlasic trend where total vertical displacement exceeds 5 km. The climate of the study area does belong to the semiarid where annual temperature goes on the average of 18°C, whereas total rainfalls are approximately 500 mm. The estimated population of the city is around 20,837. Local activities rely in a big part on agricultural activities; around 300,000 ha are dedicated to agriculture where about 1.4 million quintals are gained yearly [8, 9].

2.4 Results

2.4.1 Yearly data: the case of Mellegue catchment

2.4.1.1 ET_0

The results of polynomial regression (**Figure 7** and **Table 1**) are showing a general decrease in ET_0 compared with the initial states, which recorded 1250 mm/

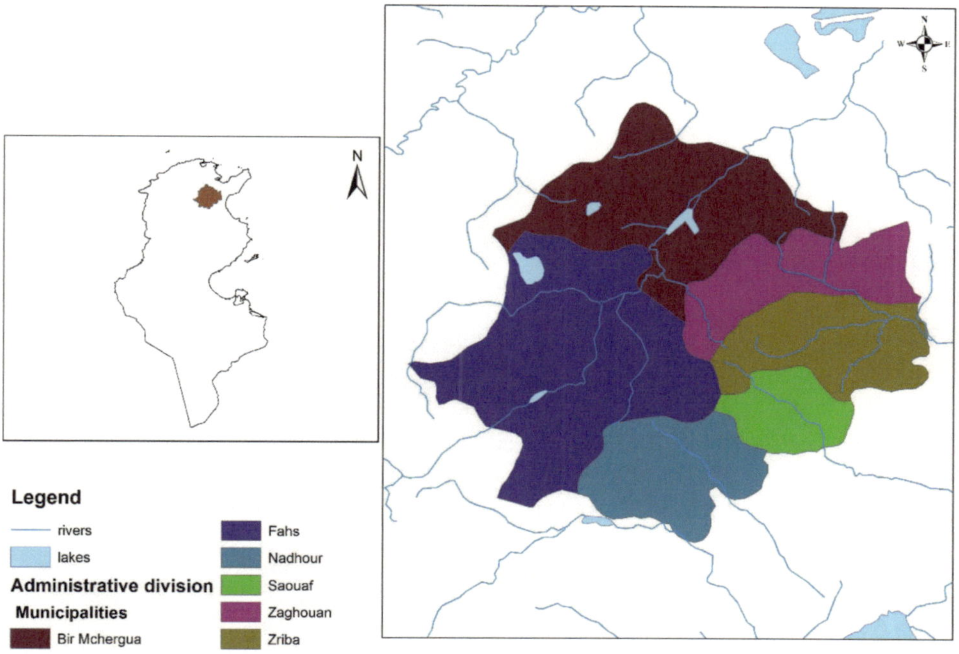

Figure 6.
Geographic location of the city of Zaghouan.

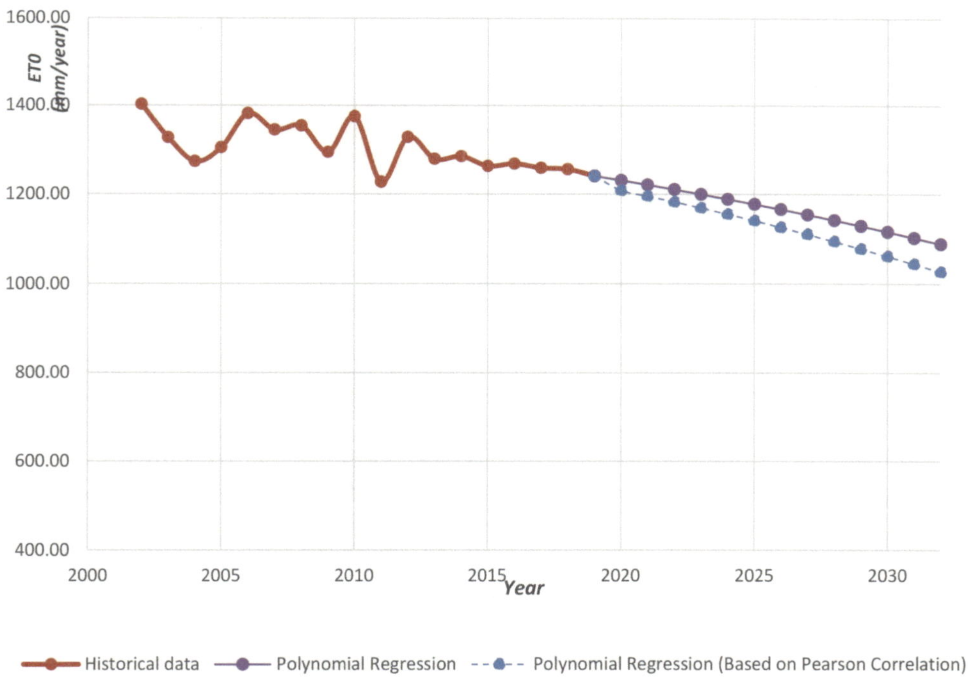

Figure 7.
Quadratic polynomial regression forecast.

year in 2019. The overall decrease is estimated at the average of 11 mm/year as it will reach, by the end of 2030, 1115 mm /year. The polynomial regression based on Pearson correlation has intensely decreasing the results, where forecasting ET_0 is estimated to reach 1060 mm in 2030 with an average loss of 16.2 mm/year. Many

Station	Poly (° = 2)				Pear Corr
	R^2	Alpha	Beta 1	Beta 2	
1	0.442	71982.188	−67.603	0.016	Valid
2		−135589.162	140.694	−0.036	Valid
3		−898739.862	903.865	−0.227	Valid
4		−1260271.181	1264.865	−0.317	Valid
5		−1060164.556	1063.572	−0.266	Valid
6		−182580.374	185.367	−0.047	Valid
7		−355144.733	358.103	−0.090	Valid
8		−333614.344	336.200	−0.084	Valid
9		−306578.982	308.705	−0.077	Valid
10		−856433.461	858.485	−0.215	Valid
11		−1394177.144	1402.250	−0.352	Valid
12		−1436810.518	1445.436	−0.363	Valid
13		−1245118.628	1252.851	−0.315	Valid
14		−878430.329	882.910	−0.222	Valid
15		−520073.866	522.075	−0.131	Valid
16		−2341253.349	2342.237	−0.585	Valid
17		−2708961.470	2715.734	−0.680	Valid
18		−856517.732	873.275	−0.222	Valid
19		−36331.810	37.729	−0.010	Valid
20		−1060164.556	1063.572	−0.266	Valid
21		2746547.110	−2734.734	0.681	Invalid
22		316197.814	−327.556	0.085	Invalid

Table 1.
Coefficient of prediction of ETo for Mellegue catchment.

causes have been mentioned earlier to link the decrease of reference evapotranspiration. But two of the most influenced factors are revealed to be the temperature and winds. In fact, many studies have predicted the increase in temperature, which may go up to 5°C (depending on the coordinates, local climate, etc.). Inversely, the rise of the surface temperature will lead to a general decrease in ET_0. This phenomenon happened to be known as the "Evaporation Paradox" [10, 11]. On the other hand, the expansion of urbanized areas will also affect ET_0 where it will undoubtedly generate more polluted air, which has a severe negative impact on the behavior of ET_0 [12]. Additionally, the general decrease of water bodies could lead to an indirect decrease in the annual value of ET_0 [13].

2.4.1.2 P

Based on the following results (**Figure 8** and **Table 2**), the forecast of rainfalls is generally stabilized with a slightly increasing tendency, where it predicts a value of precipitation that reaches 575 mm/year in 2030. Inversely, the application of Pearson polynomial regression has shown a decreasing in P with a total rainfall of 352 mm annually by the end of 2030. Based on those values, we can conclude that

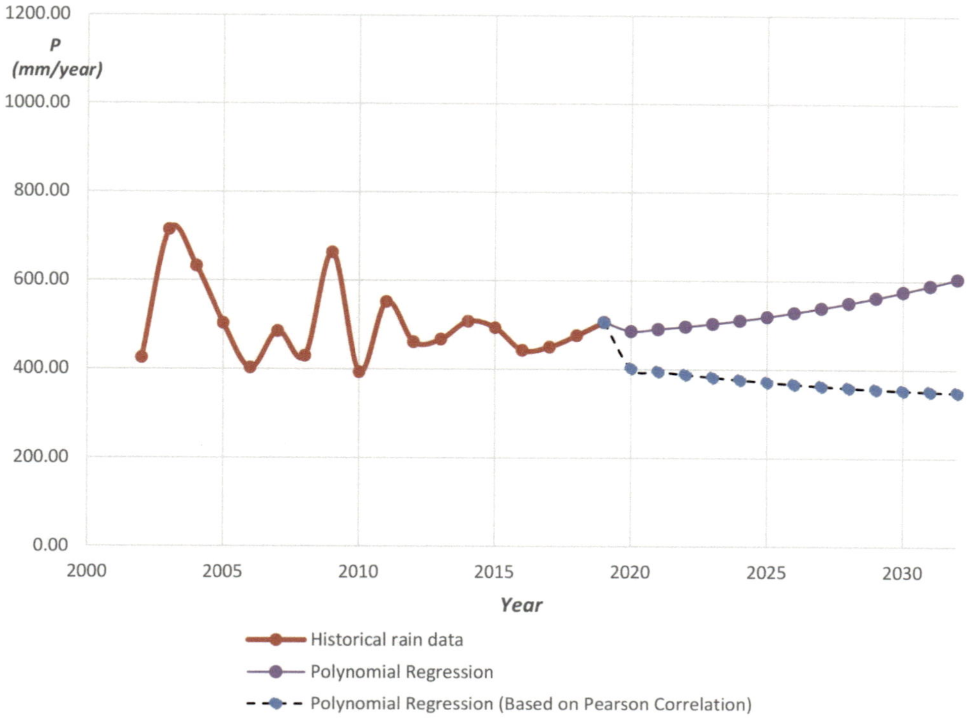

Figure 8.
Quadratic polynomial regression forecast of P.

Station	Poly (° = 2)				Pear Corr
	R^2	Alpha	Beta 1	Beta 2	
1	0.45	1955249.056	−1944.213	0.483	Invalid
2		613267.850	−612.659	0.153	Invalid
3		715018.701	−716.287	0.179	Invalid
4		2811409.072	−2808.597	0.702	Invalid
5		−362350.042	383.213	−0.101	Valid
6		2694179.641	−2670.272	0.662	Valid
7		2680735.502	−2657.038	0.658	Valid
8		2108146.344	−2084.830	0.515	Valid
9		572407.085	−558.133	0.136	Valid
10		315977.725	−312.109	0.077	Valid
11		2445104.816	−2433.165	0.605	Invalid
12		3188164.611	−3170.534	0.788	Valid
13		2497308.704	−2475.376	0.614	Valid
14		1209763.715	−1185.469	0.290	Valid
15		1464559.714	−1439.674	0.354	Valid
16		−60450.995	72.904	−0.021	Valid
17		−1642843.790	1643.988	−0.411	Valid
18		6646561.474	−6631.849	1.654	Invalid

Station	Poly (° = 2)				Pear Corr
	R^2	Alpha	Beta 1	Beta 2	
19		8793045.164	−8760.092	2.182	Invalid
20		−1108225.764	1126.116	−0.286	Valid
21		44922.783	−33.671	0.006	Valid
22		4539127.186	−4520.044	1.125	Invalid

Table 2.
Coefficient of prediction of P for Mellegue catchment.

we are still under the semiarid climate conditions. Hence, it is always too hard to predict the rainfalls because of the variety of the climate and the complexity of the related parameters (evapotranspiration, diurnal temperature, wind, relative humidity, etc.). The polynomial rise in rainfall could be explained by the general aspect of semiarid climate where precipitations are seasonally variable in intensity and quantity. Inversely, the prediction of Pearson polynomial regression agrees with many other studies that suggest a considerable loss of precipitation amounts by the year 2100, which will be noticeably remarkable in Mediterranean regions [14].

2.4.1.3 T_{max} *and* T_{min}

According to the following result, polynomial forecasts have announced a slight decrease in maximum temperature (T_{max}) of approximately 1°C compared with the initial value of 22.5°c, which was recorded in 2019 (**Figure 9** and **Table 3**). The Pearson polynomial regression has shown an opposite behavior, claiming a potential rise in T_{max}, of about 4°C (26.3°C in 2030) compared with the 2019 value. As for T_{min} (**Figure 10** and **Table 4**), the general trend is badly declining where the minimum temperature is assumed to achieve 6.7°C in 2030 where it was 8.6°C in 2019. Regarding the r polynomial regression, constructed value has shown a small

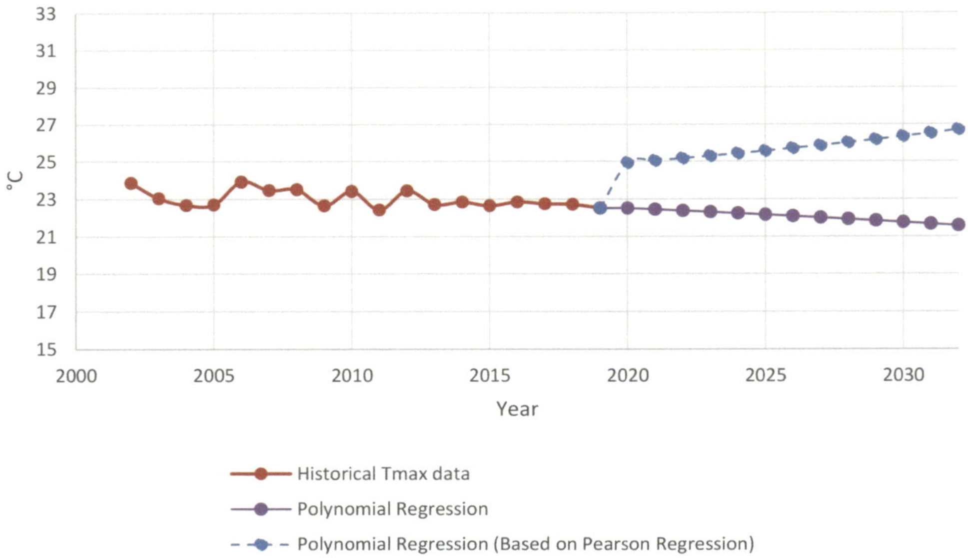

Figure 9.
Quadratic polynomial regression forecast of T_{max}.

Station	Poly (° = 2)				Pear Corr
	R^2	Alpha	Beta 1	Beta 2	
1	0.467	−4980.173	5.042	−0.001	Invalid
2		−9726.882	9.776	−0.002	Invalid
3		−14511.587	14.568	−0.004	Invalid
4		−23027.041	23.057	−0.006	Invalid
5		−2610.073	2.607	−0.001	Valid
6		−1249.637	1.281	0.000	Invalid
7		31420.577	−31.295	0.008	Valid
8		5659.772	−5.656	0.001	Valid
9		4127.832	−4.116	0.001	Valid
10		−29400.171	29.353	−0.007	Invalid
11		−20620.933	20.694	−0.005	Invalid
12		−18063.359	18.138	−0.005	Invalid
13		−12779.390	12.826	−0.003	Invalid
14		35242.347	−35.089	0.009	Valid
15		−499.410	0.506	0.000	Valid
16		−15299.754	15.350	−0.004	Invalid
17		−21567.206	21.631	−0.005	Invalid
18		−17724.813	17.840	−0.004	Invalid
19		−17130.249	17.181	−0.004	Invalid
20		−2610.073	2.607	−0.001	Valid
21		63652.590	−63.429	0.016	Valid
22		−20955.317	20.796	−0.005	Valid

Table 3.
Coefficient of prediction of Tmax for Mellegue catchment.

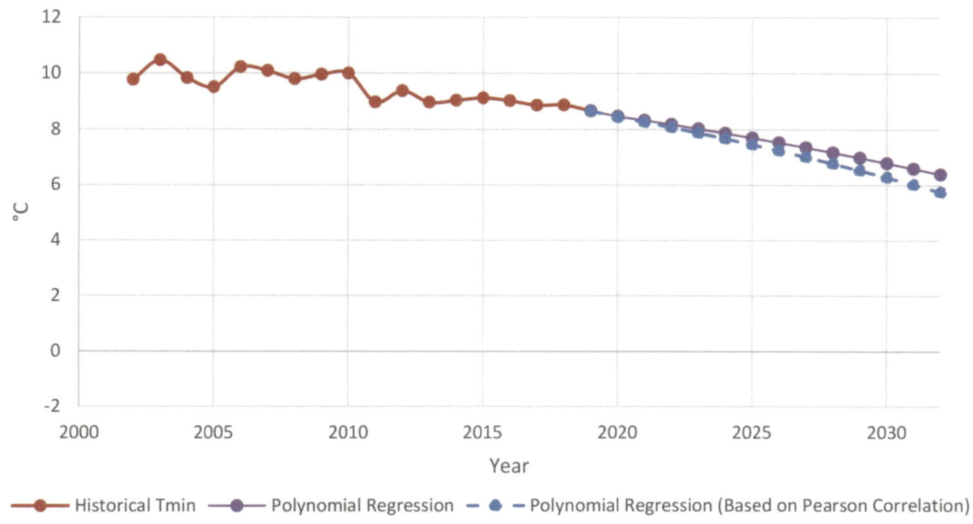

Figure 10.
Quadratic polynomial regression forecast of T_{min}.

Station	Poly (° = 2)				Pear Corr
	R^2	Alpha	Beta 1	Beta 2	
1	0.648	−15061.536	15.056	−0.0038	Valid
2		−15104.129	15.116	−0.0038	Valid
3		−16539.754	16.569	−0.0041	Valid
4		−19188.385	19.222	−0.0048	Valid
5		−17771.726	17.806	−0.0045	Valid
6		−1198.441	1.210	−0.0003	Invalid
7		1745.406	−1.719	0.0004	Invalid
8		−15167.497	15.148	−0.0038	Valid
9		−21793.646	21.777	−0.0054	Valid
10		−11902.214	11.955	−0.0030	Valid
11		−13972.995	13.983	−0.0035	Valid
12		−11936.938	11.972	−0.0030	Valid
13		−11066.352	11.111	−0.0028	Valid
14		−22346.007	22.350	−0.0056	Valid
15		−20194.330	20.203	−0.0051	Valid
16		−31769.096	31.718	−0.0079	Valid
17		−19493.627	19.517	−0.0049	Valid
18		−17911.162	17.991	−0.0045	Valid
19		−15827.268	15.904	−0.0040	Valid
20		−17771.726	17.806	−0.0045	Valid
21		55076.342	−54.831	0.0136	Invalid
22		16150.051	−15.964	0.0039	Valid

Table 4.
Coefficient of prediction of T_{min} *for Mellegue catchment.*

decrease compared with the ordinary polynomial curve, where T_{min} will reach 6.2°C by the end of 2030. The total difference between the two curves will be greater if we extend the time interval. Except for the polynomial regression based on *r* correlation, all curves are suggesting a decline in temperature during the coming period. This idea was proven to be supported in some similar cases [15]. On the other hand, forecasted T_{max} based on Pearson polynomial regression matches the whole idea of many studies that claim a general increase in temperature [16, 17]. Another crucial phenomenon is known as "Urban Heat Island," which evokes the impact of urbanized areas in destabilizing the surface temperature [17, 18].

2.4.2 Monthly data: the case of Zaghouan

2.4.2.1 P

According to the following results of **Figure 11**, it is found out that quartic polynomial regression (order (*n*) = 4) has predicted a general decrease in precipitation with an average loss of 20 mm/ year. In general, forecasting precipitation is always disputable. Some studies suggest, based on scientific results, a potential

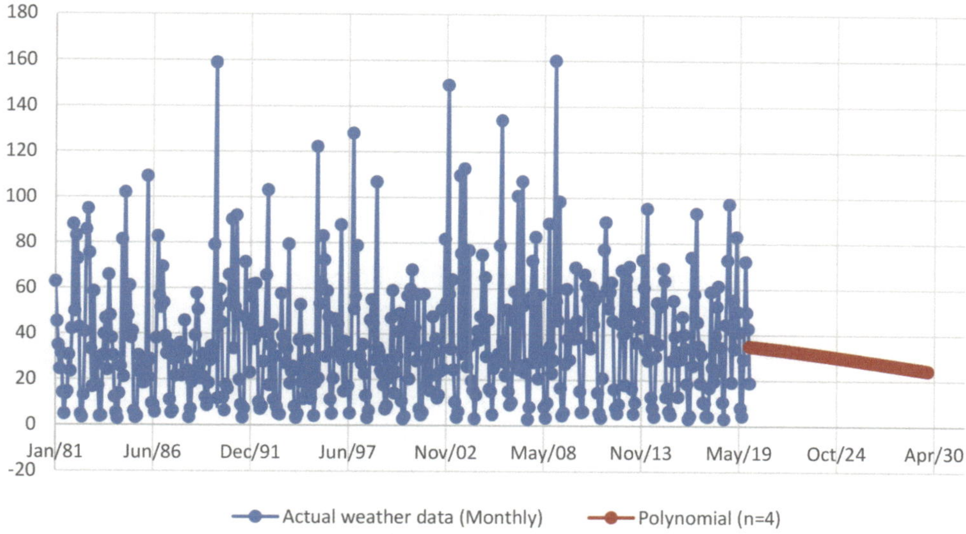

Figure 11.
Quartic polynomial regression forecast of P for the case of Zaghouan.

shortage in precipitation where the duration and intensity of drought will be intense due to an expected rise in heat. On the opposite, some places will witness a potential growth in total rainfalls due to the local climate and the geographic coordinates of the area. To summarize, we could say that warm places will become warmer leading to a shortage in precipitations, whereas wet areas will become wetter, which will intensify the amounts of rainfalls. Going back to our model, the generated trend matches the general idea that assumes a loss in forecasted rainfalls near the equator where the expected total loss of the Mediterranean areas will be around 20% [14].

3. Discussion

Considering the application of polynomial regression for the case of Mellegue watershed, it seems that the best polynomial equation is quadratic ($n = 2$). The average R^2 for the polynomial regression is around 0.5; this value happened to rise simultaneously with the advance of polynomial degree [19]. Nonetheless, the expected results are going too far from reality.

It seems that polynomial regression cannot deal with big data. The catchment is a very large basin. So, trying to understand/study a specific phenomenon needs to collect a large amount of data that has to be taken from different stations, in such a way to cover the entire area. Homogenizing the data for a resultant average response is very helpful for the machine learning algorithm to reduce error uncertainties from voluminous data, and so polynomial regression can fit the generated weather curvature [20]. Unfortunately, these data are not efficient at all for decision-makers and environmentalists because it can generate a misinterpretation as a response for treating/dealing with the potential causes for such a phenomenon.

As we see in the previous example, choosing the right order that fits mostly with the data type is often challenging to find. This is considered as one of the major handicaps in working with polynomial regression [19]. To deal with such drawbacks, choosing the right order must not be arbitrary and needs some visualization and expert of dealing with similar data (Plot, R^2, etc.). Additionally, going up in the order does generally rise the R^2 coefficient, but it has a negative impact in overestimating and amplifying the forecasted results. Generally, frequently used approach consists of progressively increasing the order until it successfully fits the data. We can rely on many other famous techniques in building strategies, which are the forward selection procedure and the backward elimination [21]. Despite the application of these strategies, polynomial regression is still very sensitive to the outliers and has sometimes unanticipated turns in inappropriate directions [22]. Such an example was very clear in predicting the minimum and maximum temperatures where the forecasted models (except of the Pearson polynomial regression) showed a decrease in temperatures, which was oppose to most of the studies, indicating a general increase in heat because of global warming. This is maybe due to an incorrect interference in interpolation and/or extrapolation of the original data because of the natural instability of the climate data. This sensitivity can be also due to the fewer techniques (lack) in validating the detection of the outliers. Additionally, the existence of few outliers (one or two) in the data can badly affect the results of the nonlinear analysis [20].

Another important drawback that should be reported is the incapacity of the polynomial regression to deal with very sensitive data such as weather. Even high orders cannot keep up with original data curves. Therefore, many samplers (points) will not be taken into account by the polynomial model, which will affect the projection of forecasted weather results. Additionally, as seen in most of the previous examples, we can say that the prediction of the weather shows a linear trend that does not conform to the general feature of the weather as it behaves seasonally, and so it must have a certain sinusoidal tendency. In the two previous study areas, most of the polynomial orders are quadratic equations (n = 2). This interpretation coincides with the general assumption that the first and second order of polynomial regression are mostly used [21].

Referring to the case of Zaghouan, statistics of polynomial regression have revealed a pessimistic result. Regardless of the order (n), R^2 was still too low (close to 0.1). This tells us the incapacity of polynomial regression to fit most of the 468-input data (from 1981 to 2019), and for that reason, it seems irrelevant to change the order of the equation. Accordingly, we convert the previous data to an annual form. As a result, the model corresponded differently. As shown in **Figure 9**, the behavior of the fourth polynomial order equation is found to be different from monthly to yearly data. We can distinguish that model performed more easily with

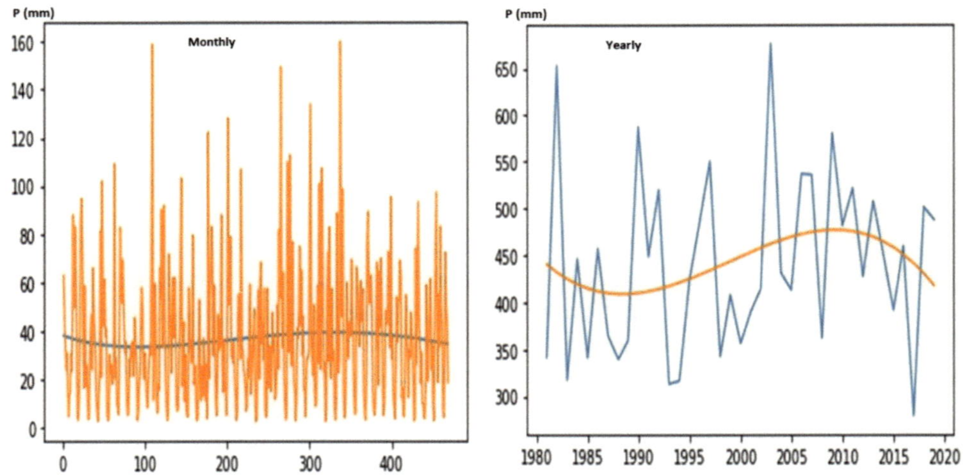

Figure 12.
Difference in the behavior of polynomial regression for monthly and yearly data (case of Zaghouan).

annual data as the R^2 has been rising to 0.4 compared with the monthly data (0.01) (**Figure 12**).

There are many advantages from using polynomial regression. Besides their simplicity in processing, they operate well in giving the best relationship between independent and dependent variables. Another good aspect regarding polynomial functions is their aptitude of fitting numerous curvatures, which depend mainly on the type and trend of the data. To conclude with, a polynomial function may not be so accurate, but it can generate an acceptable weather forecast. Let us say that it could be helpful to forecast the general behavior of climate at a certain specific range of time without the need of being too precise.

The Pearson correlation is revealed to be effective in evaluating the statistical relationship or the dependencies between two variables. Even though the method is based on the covariance and mathematical equation, it still not scientifically reliable to determine the strength and direction of the association based on Pearson correlation only. We see in the previous example of Mellegue catchment how the r coefficient has showed the strength of correlation between the stations. Every climate parameter has responded differently from the other. We can see, from **Tables 1-4** that the correlation result of a certain climate parameter is totally different from the other (such as **Tables 3** and **4** for $T_{\max}$ and $T_{\min}$ correlation). As a results, non-correlated values (less than 0.6) have been eliminated. We can see that the behavior of the prediction model based on r correlation has changed, compared with the ordinary polynomial model. Often, the results could not be notified like $T_{\min}$ cases (**Figure 10**) where there were slight changes. But it can inversely transform the curve as it is shown in **Figure 8** (rainfalls) and **Figure 9** ($T_{\max}$). In fact, the concept of using r correlation is doubtful for this case. The correlation between stations is based on the real observed values where they described the local climate of the area (temperature, wind, precipitation, etc.). We can see how big the study area is, so it is quite normal that every station will have a specific climate behavior. In other way, station records must be heterogeneous where each station will refer to the local climate of belonging area. So, by eliminating some stations based on the r coefficient, we will alter the general behavior of the catchment as a response to a certain phenomenon (land use changes, erosion, etc.). Still, r correlation can be too useful for a weather climate by evaluating the connection between two weather parameters (under the same station or area) to study the connection of one to the other. We can use it for one restricted study area (same station, for example), which had a specific local climate that has a multi-data about weather parameters. In general, the judgment of the application of Pearson correlation will mainly depend on the study example, but it is recommended to avoid their use for too heterogeneous and/or sensitive data.

4. Summary and further works

Polynomial regression algorithm was outperformed by Python for weather forecasting. We want to see the performance of the model by taking into account the type of input data where it was operated on two concrete case studies. Eventually, we want to test the relationship between stations based on Pearson correlation as a way to retrieve the data and homogenizing it. The results have opened a wide debate to discuss; convenient polynomial order was revealed to be quadratic, which agreed with the general idea that most applied polynomial order is first and second degree [21]. Additionally, the model found a good capacity to fit various complex data [22]. On the other hand, the performance of polynomial regression based on Pearson correlation has altered significantly the weather prognosis accuracy. The behavior of the model has shown a drastic change by going from monthly to yearly data. Depending on the plot observation and coefficients variables, polynomial regression happened to fit more with yearly data. This is due to the simple fact that polynomial regression will operate more efficiently with moderate to low input data. The second fact is that this type of regression is too sensitive to mutable data such as seasonal climate. That circumstance was discovered in the case of Zaghouan (monthly data), where the model has not succeeded in finding the best polynomial order (R^2 is too low). All these pieces of evidence may point to the insufficiency of polynomial algorithms in forecasting weather based on both monthly and yearly data or maybe due to the wrong conception of the model. Regardless of the reason, it is compulsory to look out for other alternative algorithms such as seasonal autoregressive integrated models (SARIMA), which can deal more efficiently with voluminous and unstable data. Also, we should improve the confidence of the inputs data by applying other machine learning algorithms as a way to homogenize the data and eliminate uncorrelated samples. However, this would require much more processing time and more complex algorithms, which will be deferred to future work.

Author details

Okba Weslati*, Samir Bouaziz and Mohamed Moncef Serbaji
Laboratory of Water, Environment, and Energy, National Engineering School of Sfax, University of Sfax, Sfax, Tunisia

*Address all correspondence to: okba.weslati@gmail.com

References

[1] National Research Council. When Weather Matters. Washington, D.C., United States: National Academy of Sciences; 2010

[2] Mavromatidis LE, Bykalyuk A, Lequay H. Development of polynomial regression models for composite dynamic envelopes' thermal performance forecasting. Applied Energy. 2013;**104**:379-391

[3] Zjavka L. Wind speed forecast correction models using polynomial neural networks. Renewable Energy. 2015;**83**:998-1006

[4] Bradley RA, Srivastava SS. Correlation in polynomial regression. The American Statistician. 1979;**33**(1):11-14

[5] Ostertagová E. Modelling using polynomial regression. Procedia Engineering. 2012;**48**:500-506

[6] McQuistan A. Using Machine Learning to Predict the Weather: Part 2 [Internet]. stackabuse.com. 2017. Available from: https://stackabuse.com/using-machine-learning-to-predict-the-weather-part-2/?fbclid=IwAR3LAOyad J48EMsG5NuTS0dKJIuLCjoGc 8YmqJmP0SNIZ0hdtTIkyIpLr40

[7] Rodier JA, Colombani J, Claude J, Kallel R. Le Bassin de la Mejerdah [Internet]. 1981. 451 p. Available from: https://www.worldcat.org/title/bassin-de-la-mejerdah/oclc/469086740

[8] Ameur M. Hamzaoui–Azaza F, Gueddari M. nitrate contamination of Sminja aquifer groundwater in Zaghouan, Northeast Tunisia: WQI and GIS assessments. Desalination and Water Treatment. 2016;**57**(50): 23698-23708

[9] Souissi F, Jemmali N, Souissi R, Dandurand JL. REE and isotope (Sr, S, and Pb) geochemistry to constrain the genesis and timing of the F-(Ba-Pb-Zn) ores of the Zaghouan District (NE Tunisia). Ore Geology Reviews [internet]. 2013; **55**(C):1-12. DOI: 10.1016/j.oregeorev.2013.04.001

[10] Roderick ML, Farquhar GD. The Cause of Decreased Pan Evaporation over the past 50 Years. Science (80-). 2002;**298**(5597):1410-1411

[11] Lin P. Impacts of climate change on reference evapotranspiration in the Qilian Mountains of China: Historical trends and projected changes. International Journal of Climatology. 2018; **38**(7):1-14

[12] Yao L. Causative impact of air pollution on evapotranspiration in the North China plain. Environmental Research [internet]. 2017;**158**:436-442. DOI: 10.1016/j.envres.2017.07.007

[13] Ramarohetra J, Sultan B. Impact of ET0 Method on the Simulation of Historical and Future Crop Yields: A Case Study of Millet Growth in Senegal. International Journal of Climatology. 2017; **38**(2):729–741

[14] Hausfather Z. Explainer: What climate models tell us about future rainfall. Carbon Brief [Internet]. 2018; Available from: https://www.carbonbrief.org/explainer-what-climate-models-tell-us-about-future-rainfall

[15] Bathiany S, Dakos V, Scheffer M, Lenton TM. Climate models predict increasing temperature variability in poor countries. Science Advances. 2018; **4**(5):1-11

[16] Lapenis A. A 50-Year-Old Global Warming Forecast That Still Holds Up. Eos [Internet]. 2020. DOI: 10.1029/2020EO151822

[17] Ackerman B. Temporal march of the Chicago heat island. Journal of Applied

Meteorology and Climatology. 1985; **24**(6):547-554

[18] Zhang J, Dong W, Wu L, Wei J, Chen P, Lee DK. Impact of land use changes on surface warming in China. Advances in Atmospheric Sciences. 2005;**22**(3):343-348

[19] Edwards JR, Parry ME. On the use of polynomial regression equations as an alternative to difference scores in organizational research. Academy of Management Journal. 1993;**36**(6): 1577-1613

[20] Zaw WT, Naing TT. Modeling of rainfall prediction over Myanmar using polynomial regression. Proc - 2009 Int Conf Comput Eng Technol ICCET 2009. 2009;**1**:316-320

[21] Shalabh IK. Polynomial regression models. In: Regression Analysis. Kanpur: Indian Institute of Technology Kanpur; 2012. pp. 1-12

[22] Qiu S, Li S, Wang F, Wen Y, Li Z, Li Z, et al. An energy exchange efficiency prediction approach based on multivariate polynomial regression for membrane-based air-to-air energy recovery ventilator core. Building and Environment. 2019;**149**(December 2018):490-500. DOI: 10.1016/ j.buildenv.2018.12.052

Chapter 8

Using Shifted Jacobi Polynomials to Handle Boundary Value Problems of Fractional Order

Kamal Shah, Eiman, Hammad Khalil, Rahmat Ali Khan and Thabet Abdeljawad

Abstract

This paper is concerned about the study of shifted Jacobi polynomials. By means of these polynomials, we construct some operational matrices of fractional order integration and differentiations. Based on these matrices, we develop a numerical scheme for the boundary value problems of fractional order differential equations. The construction of the procedure is new one for the coupled systems of fractional order boundary value problems. In the proposed scheme, we obtain a simple but highly accurate systems of algebraic equations. These systems are easily soluble by means of Matlab or using Mathematica. We provide some examples to which the procedure is applied to verify the applicability of our proposed method.

Keywords: Jacobi polynomials, operational matrices, fractional order differential equations, coupled system, boundary conditions

1. Introduction

Recently, fractional order differential equations have gained much attention from the researchers of mathematics, physics, computer science and engineers. This is due to the large numbers of applications of fractional order differential equations in various fields of science and applied nature. The applications of fractional order differential equations are found in physics, mechanics, viscoelasticity, photography, biology, chemistry, fluid mechanics, image and signal processing phenomenons, etc.; for more details, see [1–4]. The researchers are giving more attention to fractional order differential equations, because in most cases the fractional order models are more accurate and reliable than classical order models. On the other hand, fractional order models have more degrees of freedom than integer order models. In most cases, the hereditary phenomena and memory description process are accurately modeled by means of fractional order derivatives and integrations as compare with integer order derivative and integrations. Due to these facts, researchers had given much attention to study the existence and uniqueness of positive solutions as well as multiplicity of solutions for fractional order differential equations and their systems. This area is very well explored, and large numbers of research articles can be found in literature; for details, see [5–10]. It is very difficult to obtain exact solutions for such type of differential equations in all cases always. However, the area involving numerical solutions is in initial stage. Therefore, various numerical schemes have been developed in last few years by different researchers. The aim of these schemes was to achieve better accuracy. These schemes have their own merits and demerits. For example, in [11–13], the authors developed Adomain decomposition methods, Homotopy analysis methods, and Variational iterational methods to obtained good approximate solutions for certain fractional order differential equations. In recent years, spectral methods have got much attention as they were applied to solve some real-word problems of various fields of science and engineering. High accuracy was obtained for solving such problems [14]. Spectral method needs operational matrices for the numerical solutions, which have been constructed by using some polynomials, for example, in [15], the authors developed an operational matrix for shifted Legendre polynomials corresponding to fractional order derivative. In [16], the authors introduced operational matrix by using shifted Legendre polynomials corresponding to fractional order integrations. Similarly in [17, 18], authors constructed operational matrices for fractional order derivative by using Chebyshev and Jacobi polynomials. In all these cases, these matrices were applied to solve multiterms fractional order differential equations together with Tau-collocation method. In [19], Singh et al. derived an operational matrices for Berestein polynomials corresponding to fractional order integrations. Haar wavelet operational matrices were also developed, and some problems of fractional order differential equations were solved. In all these methods, collocation method was used together with these methods to obtain numerical solutions for fractional order differential equations. They only solved single problems.

Most of the physical and biological phenomena can be modeled in the form of coupled systems. For example, when two masses and two spring systems are modeled, it led to a coupled system of differential equations.

$$\mathrm{D}^{\alpha}X+\frac{\lambda}{m_1}\mathrm{D}^{\nu_1}X-\frac{\mu}{m_1}\mathrm{D}^{\nu_2}Y+\frac{\kappa_1}{m_1}X-\frac{\kappa_2}{m_1}Y=f(x),\quad 1<\alpha\leq 2,\quad 0<\nu_1,\nu_2\leq 1,$$
$$\mathrm{D}^{\beta}Y-\frac{\lambda}{m_2}\mathrm{D}^{\omega_1}X+\frac{\lambda+\mu}{m_2}\mathrm{D}^{\omega_2}Y-\frac{\kappa_1}{m_2}X+\frac{\kappa_1+\kappa_2}{m_2}Y=g(x),\quad 1<\beta\leq 2,\quad 0<\omega_1,\omega_2\leq 1,$$

subject to the boundary conditions

$$X(0)=0,\quad X(1)=0$$
$$Y(0)=1,\quad Y(1)=-1.$$

Many coupled systems can be found in physics and in biological model, chemistry, fluid dynamics, etc.; see [16, 20–27]. In fractional calculus, we generalize these systems by replacing the integer order by any fractional values that lie in certain range. All the above numerical methods were applied to solve singles problems, and to the best of our knowledge, very few articles are devoted to the numerical solutions of coupled systems or models. In all these cases, the authors have solved initial value problems of coupled systems of fractional order differential and partial differential equations; for details, see [15, 28] and the references therein. For example, in [15], authors have solved some initial value problems of coupled systems of fractional differential equations (FDEs) by using shifted Jacobi polynomials operational matrix method. However, the scheme is not properly applied yet for solving coupled systems of boundary value problems of fractional order differential equations. In this paper, we discuss the shifted Jacobi polynomial operational matrices methods to solve boundary value problems for a coupled systems of fractional order differential equations. In this scheme, we introduced a new matrix corresponding to boundary conditions, which is required for the approximate solutions. First, we solve single boundary value problems given by

$$\mathrm{D}^{\alpha}x(t) + A\mathrm{D}^{\nu}x(t) + B\mathrm{D}^{\omega}x(t) + Cx(t) = f(t), t \in [0, T], A, B, C \in \mathbb{R}, \tag{1}$$

subject the boundary conditions

$$x(0) = x_0, \quad x(T) = x_1,$$

where $f(t) \in C[0, T]$ is a source term and $1 < \alpha \leq 2, \; 0 < \nu, \omega < 1$. Then, we extend our technique to solve coupled system of fractional differential equations (FDEs) given as

$$\begin{aligned} &\mathrm{D}^{\alpha}x(t) + A_1\mathrm{D}^{\nu_1}x(t) + B_1\mathrm{D}^{\nu_2}y(t) + C_1x(t) + D_1y(t) = f(t), \;\; t \in [0, T], \\ &\mathrm{D}^{\beta}y(t) + A_2\mathrm{D}^{\omega_1}x(t) + B_2\mathrm{D}^{\omega_2}y(t) + C_2x(t) + D_2y(t) = g(t), \;\; t \in [0, T], \end{aligned} \tag{2}$$

subject to the boundary conditions given by

$$\begin{aligned} x(0) = x_0, \;\; x(T) = x_1, \\ y(0) = y_0, \;\; y(T) = y_1, \end{aligned} \tag{3}$$

where $1 < \alpha, \beta \leq 2, \;\; 0 < \nu_i, \omega_i < 1 (i = 1, 2)$. Further, $f(t), g(t) \in C[0, T]$ are the source term of the system (42), (43) and x_0, y_0, x_1, y_1 are real constants, $A_i, B_i, C_i, D_i (i = 1, 2)$ are any constants. We convert the differential equations/systems of differential equations to a coupled systems of simple algebraic equations of Selvester or Laypunove type, which are easily soluble for unknown matrix required for approximations. In our proposed method, we do no need collation method, which is required for abovementioned methods. We applied our technique to some practical problems to verify the applicability of the proposed method.

Our article is organized as follows: Some fundamental concepts and results of fractional calculus and Jacobi polynomials needed are provided in sections 2. Section 3 contains some operational matrices for shifted Jacobi polynomials corresponding to fractional order derivative and integrations. Also in this section, we provide new operational matrix corresponding to the boundary conditions. In Section 4, we derive main procedure for the general class of boundary value problems of fractional order differential equations as well as for coupled system of FDEs. In Section 5, we apply the proposed method to some examples. Also we plot the images of these examples. In Section 6, we provide a short conclusion of the papers.

2. Preliminaries

In this section, we recall some basic definitions and results of fractional calculus, also we will give fundamental properties, definitions related to Jacobi polynomials as in [1, 4, 6–8, 14, 17, 18, 28–31].

Definition 2.1. The fractional integral of order $\alpha > 0$ of a function $z : (0, \infty) \to R$ is defined by

$$\mathbf{I}^{\alpha}z(t) = \frac{1}{\Gamma(\alpha)} \int_0^t \frac{z(s)}{(t-s)^{1-\alpha}} ds, \tag{4}$$

provided the integral converges at the right sides. Further, a simple and important property of $\mathbf{I}^{\alpha}$ is given by

$$\mathbf{I}^{\alpha}z^{\delta} = \frac{\Gamma(\delta+1)}{\Gamma(\delta+\alpha+1)} z^{\alpha+\delta}. \tag{5}$$

Definition 2.2. *The Caputo fractional derivative of order $\alpha > 0$ of a function $z \in C^n[0,1]$ is defined by*

$$D^{\alpha}_{0+}z(t) = \frac{1}{\Gamma(n-\alpha)}\int_0^t (t-s)^{n-\alpha-1}\frac{d^n}{dt^n}z(s)ds, \quad n-1<\alpha\leq n, t>0, \tag{6}$$

where $n = \lceil \alpha \rceil + 1$,

provided that the right side is pointwise defined on $(0,\infty)$. Also one of the important properties of fractional order derivative is given by

$$D^{\alpha}t^k = \frac{\Gamma(k+1)}{\Gamma(k-\alpha+1)}t^{k-\alpha}. \text{ Also for any cons tan t } C, \text{we have } D^{\alpha}C = 0. \tag{7}$$

The following results are needed in the sequel.
Lemma 2.2.1. [6], Let $\alpha > 0$ then

$$I^{\alpha}D^{\alpha}y(t) = y(t) + C_0 + C_1 t + \ldots + C_{n-1}t^{\alpha-n}, \tag{8}$$

for arbitrary

$$C_i \in \mathbb{R}, \ i = 0,1,2,\ldots,n-1, \ n = [\alpha]+1.$$

2.1 The shifted Jacobi polynomials and its fundamental properties

In this section, we provide basic properties of shifted Jacobi polynomials. The famous Jacobi polynomials $P_i^{(a,b)}(z)$ are defined over the interval $[-1,1]$ as given by

$$P_i^{(a,b)}(z) = \frac{(a+b+2i-1)\left[a^2-b^2+z(a+b+2i-2)(a+b+2i-2)\right]}{2i(a+b+i)(a+b+2i-2)}P_{i-1}^{(a,b)}(z)$$
$$-\frac{(a+i-1)(b+i-1)(a+b+2i)}{i(a+b+i)(a+b+2i-2)}P_{i-2}^{(a,b)}(z), \quad i=2,3, \ \ldots,$$
$$\text{where} \quad P_0^{(a,b)}(z)=1, \ P_1^{(a,b)}(z) = \frac{a+b+2}{2}z+\frac{a-b}{2}. \tag{9}$$

By means of the substitution $z = \frac{2t}{\xi} - 1$, we get a new version of the above polynomials defined in (9), which is called the shifted Jacobi polynomials over the interval $[0,\xi]$. The analytical form of this shifted Jacobi polynomials is given by the relations

$$P_{\xi,i}^{(a,b)}(t) = \sum_{k=0}^{i}\frac{(-1)^{i-k}\Gamma(i+b+1)\Gamma(i+k+a+b+1)}{\Gamma(k+b+1)\Gamma(i+a+b+1)\Gamma(i-k+1)\Gamma(k+1)\xi^k}t^k,$$
$$\text{where } P_{\xi,i}^{(a,b)}(0) = (-1)^i\frac{\Gamma(i+b+1)}{\Gamma(b+1)\Gamma(i+1)}. \tag{10}$$

The orthogonality conditions of the shifted Jacobi polynomials are given by

$$\int_0^{\xi} P_{\xi,j}^{(a,b)}(t)P_{\xi,i}^{(a,b)}(t)W_{\xi}^{(a,b)}(t)dt = \Omega_{\xi,j}^{(a,b)}\delta_{ji},$$
$$\text{where } \delta_{ji}=1, \quad \text{if } i=j, \quad \text{other wise } \delta_{ji}=0, \tag{11}$$
$$\text{and the weight function is given by } W_{\xi}^{(a,b)}(t) = (\xi-t)^a t^b,$$

and

$$\Omega_{\xi,j}^{(a,b)}(t) = \frac{\xi^{a+b+1}\Gamma(j+a+1)\Gamma(j+b+1)}{(2j+a+b+1)\Gamma(j+1)\Gamma(j+a+b+1)}. \tag{12}$$

Any function $f(t)$ such that $f^2(t)$ is integrable over $[0, \xi]$, can be approximated in terms of shifted Jacobi polynomials as given by

$$f(t) \simeq \sum_{k=0}^{m} B_j P_{\xi,k}^{(a,b)}(t) = \mathrm{K}_M^T \Phi_M(t), \tag{13}$$

where the shifted Jacobi coefficient vector is denoted by K_M and $\Phi_M(t)$ is M terms function vector. Also $M = m + 1$, when $m \to \infty$ the approximation converges to the exact value of the function. The coefficient b_k can be calculated by using (11)–(13) as

$$B_j = \frac{1}{\Omega_{\xi,j}^{(a,b)}} \int_0^{\xi} W_{\xi}^{(a,b)}(t) f(t) P_{\xi,i}^{(a,b)}(t) dt, \ \ i = 0, 1, \ldots. \tag{14}$$

3. Operational matrices

In this section, we provide some operational matrices for fractional order differentiation as well as for fractional order integration. Further, in this section, we give construction of new matrix for the boundary conditions.

Theorem 3.1. Let $\Phi_M(t)$ be the function vector, then the operational matrix of fractional order derivative α is given by

$$\mathrm{D}^{\alpha}[\Phi_M(t)] \simeq \mathrm{H}_{M\times M}^{(\alpha)} \Phi_M(t), \tag{15}$$

where $\mathrm{H}_{M\times M}^{(\alpha)}$ is the operational matrix of fractional order derivative α given by

$$\mathbf{H}_{M\times M}^{(\alpha)} = \begin{bmatrix} \Upsilon_{0,0,k} & \Upsilon_{0,1,k} & \cdots & \Upsilon_{0,j,k} & \cdots & \Upsilon_{0,m,k} \\ \Upsilon_{2,1,k} & \Upsilon_{2,2,k} & \cdots & \Upsilon_{2,r,k} & \cdots & \Upsilon_{1,m,k} \\ \vdots & \vdots & \vdots & \vdots & \vdots & \vdots \\ \Upsilon_{i,0,k} & \Upsilon_{i,1,k} & \cdots & \Upsilon_{i,j,k} & \cdots & \Upsilon_{i,m,k} \\ \vdots & \vdots & \vdots & \vdots & \vdots & \vdots \\ \Upsilon_{m,0,k} & \Upsilon_{m,1,k} & \cdots & \Upsilon_{m,j,k} & \cdots & \Upsilon_{m,m,k} \end{bmatrix}, \tag{16}$$

where

$$\Upsilon_{i,j,k} = \sum_{k=0}^{i} \frac{(-1)^{i-k}\Gamma(i+b+1)\Gamma(i+k+a+b+1)\Gamma(k+1)}{\Gamma(k+b+1)\Gamma(i+a+b+1)\Gamma(i-k+1)\Gamma(k+1)\Gamma(k+1-\alpha)\xi^k}$$
$$\times \sum_{l=0}^{j} \frac{(-1)^{j-l}(2j+a+b+1)\Gamma(j+1)\Gamma(j+l+a+b+1)\Gamma(k-alpha+l+b+1)\Gamma(a+1)\xi^{\alpha}}{\Gamma(j+a+1)\Gamma(l+b+1)\Gamma(j-l+1)\Gamma(l+1)\Gamma(k-\alpha+l+b+a+2)}.$$
$$\text{If} \ \ \Upsilon_{i,j,k} = 0, \ \ \text{for} \ \ i < \alpha. \tag{17}$$

Proof. The proof of this theorem is same as given in [14, 17] in lemma (3.2) and (3.4), respectively. □

Theorem 3.2. *Let $\mathbf{\Phi}_M(t)$ be the function vector corresponding to the shifted Jacobi polynomials and $\alpha > 0$, then the operational matrix corresponding to the fractional order integration is given by*

$$\mathrm{I}^{\alpha}[\Phi_M(t)] \simeq \mathrm{H}^{(\alpha)}_{M\times M}\Phi_M(t), \tag{18}$$

where $\mathrm{H}^{(\alpha)}_{M\times M}$ is the operational matrix of fractional order integration α and is given by

$$\mathbf{H}^{(\alpha)}_{M\times M} = \begin{bmatrix} \Upsilon_{0,0,k} & \Upsilon_{0,1,k} & \cdots & \Upsilon_{0,j,k} & \cdots & \Upsilon_{0,m,k} \\ \Upsilon_{2,1,k} & \Upsilon_{2,2,k} & \cdots & \Upsilon_{2,r,k} & \cdots & \Upsilon_{1,m,k} \\ \vdots & \vdots & \vdots & \vdots & \vdots & \vdots \\ \Upsilon_{i,0,k} & \Upsilon_{i,1,k} & \cdots & \Upsilon_{i,j,k} & \cdots & \Upsilon_{i,m,k} \\ \vdots & \vdots & \vdots & \vdots & \vdots & \vdots \\ \Upsilon_{m,0,k} & \Upsilon_{m,1,k} & \cdots & \Upsilon_{m,j,k} & \cdots & \Upsilon_{m,m,k} \end{bmatrix}, \tag{19}$$

where

$$\Upsilon_{i,j,k} = \sum_{k=0}^{i} \frac{(-1)^{i-k}\Gamma(i+b+1)\Gamma(i+k+a+b+1)\Gamma(k+1)}{\Gamma(k+b+1)\Gamma(i+a+b+1)\Gamma(i-k+1)\Gamma(k+1)\Gamma(k+1+\alpha)\xi^{k}}$$
$$\times \sum_{l=0}^{j} \frac{(-1)^{j-l}(2j+a+b+1)\Gamma(j+1)\Gamma(j+l+a+b+1)\Gamma(k+\alpha+l+b+1)\Gamma(a+1)\xi^{\alpha}}{\Gamma(j+a+1)\Gamma(l+b+1)\Gamma(j-l+1)\Gamma(l+1)\Gamma(k+\alpha+l+b+a+2)}. \tag{20}$$

Proof. The proof of this theorem is available in [14]. Therefore, we omit the proof.

Theorem 3.3 *Let $\Phi_M(t)$ be a function vector, and let $\phi(t)$ be any function defined as $\phi(t) = ct^n$, $n = 0, 1, 2, \ldots, c \in \mathbb{R}$ and $u(t) = \mathrm{K}^T_M\Phi_M(t)$. Then*

$$\phi(t)\,{}_0\mathrm{I}^{\alpha}_{\xi}u(t) = \mathrm{K}^T_M \mathrm{Q}^{(c,\phi,\alpha)}_{M\times M}\Phi_M(t), \tag{21}$$

where $\mathrm{Q}^{(c,\phi,\alpha)}_{M\times M}$ *is an operational matrix corresponding to some boundary value and is given by*

$$\mathbf{Q}^{(c,\phi,\alpha)}_{M\times M} = \begin{bmatrix} \Upsilon_{0,0,k} & \Upsilon_{0,1,k} & \cdots & \Upsilon_{0,j,k} & \cdots & \Upsilon_{0,m,k} \\ \Upsilon_{2,1,k} & \Upsilon_{2,2,k} & \cdots & \Upsilon_{2,r,k} & \cdots & \Upsilon_{1,m,k} \\ \vdots & \vdots & \vdots & \vdots & \vdots & \vdots \\ \Upsilon_{i,0,k} & \Upsilon_{i,1,k} & \cdots & \Upsilon_{i,j,k} & \cdots & \Upsilon_{i,m,k} \\ \vdots & \vdots & \vdots & \vdots & \vdots & \vdots \\ \Upsilon_{m,0,k} & \Upsilon_{m,1,k} & \cdots & \Upsilon_{m,j,k} & \cdots & \Upsilon_{m,m,k} \end{bmatrix}, \tag{22}$$

where

$$\Upsilon_{i,j,k} = \sum_{j=0}^{M} \Delta_{i,k,\alpha} B_j P^{(a,b)}_{\xi,j}(t), \tag{23}$$

where

$$\Delta_{i,k,\alpha} = \sum_{k=0}^{i} \frac{(-1)^{i-k}(i+b+1)\Gamma(i+a+b+1)c\xi^{\alpha}}{\Gamma(k+b+1)\Gamma(i+a+b+1)\Gamma(i-k+1)\Gamma(k+\alpha)} \tag{24}$$

and

$$B_j = \sum_{l=0}^{j} \frac{(-1)^{j-l}(2j+a+b+1)\Gamma(j+1)\Gamma(j+l+a+b+1)\Gamma(n+b+1)\Gamma(a+1)\xi^{n-l}}{\Gamma(j+a+1)\Gamma(l+b+1)\Gamma(j-l+1)\Gamma(l+1)\Gamma(n+a+b+1)}. \tag{25}$$

Proof. Consider the general term of $\Phi_M(t)$ as

$$\begin{aligned}
{}_0\mathrm{I}_\xi^\alpha P_{\xi,i}^{(a,b)}(t) &= \frac{1}{\Gamma\alpha}\int_0^\xi (\xi-s)^{\alpha-1}P_{\xi,i}^{(a,b)}(s)ds \\
&= \frac{1}{\Gamma\alpha}\int_0^\xi (\xi-s)^{\alpha-1}\sum_{k=0}^{i}\frac{(-1)^{i-k}(i+b+1)\Gamma(i+k+a+b+1)s^k}{\Gamma(k+b+1)\Gamma(i+a+b+1)\Gamma(i-k+1)\Gamma(k+1)\xi^k}ds \\
&= \frac{1}{\Gamma\alpha}\sum_{k=0}^{i}\frac{(-1)^{i-k}(i+b+1)\Gamma(i+k+a+b+1)}{\Gamma(k+b+1)\Gamma(i+a+b+1)\Gamma(i-k+1)\Gamma(k+1)\xi^k}\int_0^\xi (\xi-s)^{\alpha-1}s^k ds.
\end{aligned} \tag{26}$$

Now by applying Laplace transform, we get

$$\begin{aligned}
&\mathcal{L}\left(\int_0^\xi (\xi-s)^{\alpha-1}s^k ds\right) = \frac{\Gamma(\alpha)\Gamma(k+1)}{s^{k+\alpha+1}} \\
&\Rightarrow \int_0^\xi (\xi-s)^{\alpha-1}s^k ds = \frac{\Gamma(k+1)\Gamma\alpha\ \xi^{k+\alpha}}{\Gamma(\alpha+1)}.
\end{aligned} \tag{27}$$

Now putting (27) in (26), we get

$${}_0\mathrm{I}_\xi^\alpha P_{\xi,i}^{(a,b)}(t) = \sum_{k=0}^{i}\frac{(-1)^{i-k}(i+b+1)\Gamma(i+a+b+k+1)\xi^\alpha}{\Gamma(k+b+1)\Gamma(i+a+b+1)\Gamma(i-k+1)\Gamma(k+\alpha)}. \tag{28}$$

Now if $\phi(t) = t^n$, then

$$c\phi(t)_0\mathrm{I}_\xi^\alpha P_{\xi,i}^{(a,b)}(t) = \sum_{k=0}^{i}\frac{(-1)^{i-k}(i+b+1)\Gamma(i+a+b+k+1)\xi^\alpha c t^n}{\Gamma(k+b+1)\Gamma(i+a+b+1)\Gamma(i-k+1)\Gamma(k+\alpha)}. \tag{29}$$

Representing t^n in terms of shifted Jacobi polynomials as

$$t^n = \sum_{j=0}^{M} B_j P_{\xi,j}^{(a,b)}(t). \tag{30}$$

By the use of orthogonality relation (11), we have

$$\begin{aligned}
B_j &= \frac{1}{\Omega_{\xi,j}^{(a,b)}}\int_0^\xi t^n P_{\xi,j}^{(a,b)}(t)W_\xi^{(a,b)}(\xi)dt \\
&= \frac{1}{\Omega_{\xi,j}^{(a,b)}}\sum_{l=0}^{j}\frac{(-1)^{(j-l)}\Gamma(j+b+1)\Gamma(j+l+a+b+1)}{\Gamma(l+b+1)\Gamma(j+a+b+1)\Gamma(j-l+1)\Gamma(l+1)\xi^l}\int_0^\xi t^{n+b}(\xi-t)^a dt.
\end{aligned} \tag{31}$$

Now by means of Laplace transform, we have

$$\mathrm{L}\left(\int_0^{\xi} t^{n+b}(\xi-t)^a dt\right) = \frac{\Gamma(n+b+1)\Gamma(a+1)}{\xi^{n+b+a+2}}$$
$$\Rightarrow \int_0^{\xi} t^{n+b}(\xi-t)^a dt = \frac{\Gamma(n+b+1)\Gamma(a+1)}{\Gamma(n+a+b+1)}\xi^{n+a+b+1}. \tag{32}$$

Thus putting (32) in (31) and using (12), we get

$$B_j = \sum_{l=0}^{j} \frac{(-1)^{j-l}(2j+a+b+1)\Gamma(j+1)\Gamma(j+l+a+b+1)\Gamma(n+b+1)\Gamma(a+1)\xi^{n-l}}{\Gamma(j+a+1)\Gamma(l+b+1)\Gamma(j-l+1)\Gamma(l+1)\Gamma(n+a+b+1)}. \tag{33}$$

Now (29) implies that

$$\phi(t)_0\mathrm{I}_{\xi}^{\alpha}P_{\xi,i}^{(a,b)}(t) = \sum_{j=0}^{M}\sum_{k=0}^{i} \frac{(-1)^{i-k}(i+b+1)\Gamma(i+a+b+k+1)\xi^{\alpha}c}{\Gamma(k+b+1)\Gamma(i+a+b+1)\Gamma(i-k+1)\Gamma(k+\alpha)} B_j P_{\xi,j}^{(a,b)}(t)$$
$$= \sum_{j=0}^{M} \Delta_{i,k,\alpha} B_j P_{\xi,j}^{(a,b)}(t) = \sum_{j=0}^{M} \Upsilon_{i,j,k} P_{\xi,j}^{(a,b)}(t), \tag{34}$$

where

$$\Upsilon_{i,j,k} = \sum_{j=0}^{M} \Delta_{i,k,\alpha} B_j,$$

$$\Delta_{i,k,\alpha} = \sum_{k=0}^{i} \frac{(-1)^{i-k}(i+b+1)\Gamma(i+a+b+k+1)\xi^{\alpha}c}{\Gamma(k+b+1)\Gamma(i+a+b+1)\Gamma(i+k+1)\Gamma(k+\alpha)}.$$

Evaluating the result (34) for different values of j we get the required result.

4. Applications of operational matrices

In this section, we show fundamental importance of operational matrices of fractional order derivative and integration. We apply them to solve some multiterms fractional order boundary value problems of fractional differential equations. Consider the following general fractional differential equation with constant coefficient and given boundary conditions

$$\mathrm{D}^{\alpha}x(t) + A\mathrm{D}^{\nu}x(t) + B\mathrm{D}^{\omega}x(t) + Cx(t) = f(t), t \in [0, T], A, B, C \in \mathbb{R}, \tag{35}$$

subject the boundary conditions

$$x(0) = x_0, \quad x(T) = x_1,$$

where $f(t) \in C[0, T]$ is a source terms and $1 < \alpha \leq 2,\ \ 0 < \nu, \omega < 1$..

To obtain the solutions of the (35) in terms of shifted Jacobi polynomials, we proceed as

$$D^{\alpha}x(t) = \mathrm{K}_M^T \Phi_M(t). \tag{36}$$

Applying I^{α}, definition (7), and lemma (2.2.1), then we get

$$x(t) = \mathrm{K}_M^T \mathrm{H}_{M\times M}^{(\alpha)} \Phi_M(t) + c_0 + c_1 t. \tag{37}$$

By means of boundary conditions we get

$$x(t) = \mathrm{K}_M^T \mathrm{H}_{M\times M}^{(\alpha)} \Phi_M(t) + x_0 + t\frac{(x_1 - x_0)}{T} - \frac{t}{T} \mathrm{K}_M^T \mathrm{H}_{M\times M}^{(\alpha)} \Phi_M(t)\Big|_{t=T} \tag{38}$$

using the approximation $x_0 + t\frac{(x_1-x_0)}{T} \simeq \mathrm{F}_M^1 \Phi_M(t)$, $\frac{t}{T} = \phi(t)$, also using $c\phi(t)\mathrm{K}_M^T \mathrm{H}_{M\times M}^{(\alpha)} \Phi_M(t)\Big|_{t=T} = \mathrm{K}_M^T \mathrm{Q}_{M\times M}^{(\alpha,\phi)} \Phi_M(t)$, then (38) implies that

$$\begin{aligned} x(t) &= \mathrm{K}_M^T \mathrm{H}_{M\times M}^{(\alpha)} \Phi_M(t) + \mathrm{F}_M^1 \Phi(t) - c\phi(t) \mathrm{K}_M^T \mathrm{H}_{M\times M}^{(\alpha)} \Phi_M(t)\Big|_{t=T} \\ &= \mathrm{K}_M^T \mathrm{H}_{M\times M}^{(\alpha)} \Phi_M(t) - \mathrm{K}_M^T \mathrm{Q}_{M\times M}^{(\alpha,\phi)} \Phi_M(t) + \mathrm{F}_M^1 \Phi_M(t) \\ &= \mathrm{K}_M^T \left(\mathrm{H}_{M\times M}^{\alpha} - \mathrm{Q}_{M\times M}^{(\alpha,\phi)}\right) \Phi_M(t) + \mathrm{F}_M^1 \Phi_M(t). \end{aligned} \tag{39}$$

Now from (39) we have

$$\begin{aligned} &\mathrm{D}^{\nu} x(t) = \mathrm{K}_M^T \left(\mathrm{H}_{M\times M}^{(\alpha)} - \mathrm{Q}_{M\times M}^{(\alpha,\phi)}\right) \mathrm{G}_{M\times M}^{(\nu)} \Phi_M(t) + \mathrm{F}_M^1 \mathrm{G}_{M\times M}^{(\nu)} \Phi_M(t) \\ &\mathrm{D}^{\omega} x(t) = \mathrm{K}_M^T \left(\mathrm{H}_{M\times M}^{(\alpha)} - \mathrm{Q}_{M\times M}^{(\alpha,\phi)}\right) \mathrm{G}_{M\times M}^{(\omega)} \Phi_M(t) + \mathrm{F}_M^1 \mathrm{G}_{M\times M}^{(\omega)} \Phi_M(t) \\ &\text{and approximating the source term as} \qquad f(t) \simeq \mathrm{F}_M^2 \Phi_M(t). \end{aligned} \tag{40}$$

Putting (36)–(40) in (35), which yields

$$\begin{aligned} &\mathrm{K}_M^T \Phi_M(t) + A\left[\mathrm{K}_M^T \left(\mathrm{H}_{M\times M}^{(\alpha)} - \mathrm{Q}_{M\times M}^{(\alpha,\phi)}\right) \mathrm{G}_{M\times M}^{(\nu)} \Phi_M(t) + \mathrm{F}_M^1 \mathrm{G}_{M\times M}^{(\nu)} \Phi_M(t)\right] \\ &+B\left[\mathrm{K}_M^T \left(\mathrm{H}_{M\times M}^{(\alpha)} - \mathrm{Q}_{M\times M}^{(\alpha,\phi)}\right) \mathrm{G}_{M\times M}^{(\omega)} \Phi_M(t) + \mathrm{F}_M^1 \mathrm{G}_{M\times M}^{(\omega)} \Phi_M(t)\right] \\ &+C\left[\mathrm{K}_M^T \left(\mathrm{H}_{M\times M}^{(\alpha)} - \mathrm{Q}_{M\times M}^{(\alpha,\phi)}\right) \Phi_M(t) + \mathrm{F}_M^1 \Phi_M(t)\right] - \mathrm{F}_M^2 \Phi_M(t) = 0 \\ &\mathrm{K}_M^T \Phi_M(t) + \mathrm{K}_M^T \left(\mathrm{H}_{M\times M}^{(\alpha)} - \mathrm{Q}_{M\times M}^{(\alpha,\phi)}\right) \left[A\mathrm{G}_{M\times M}^{(\nu)} + B\mathrm{G}_{M\times M}^{(\omega)} + CI\right] \Phi_M(t) \\ &+\left[\mathrm{F}_M^1 \left(A\mathrm{G}_{M\times M}^{(\nu)} + B\mathrm{G}_{M\times M}^{(\omega)}\right) + C\mathrm{F}_M^1 - \mathrm{F}_M^2\right] \Phi_M(t) = 0 \\ &\mathrm{K}_M^T + \mathrm{K}_M^T \left(\mathrm{H}_{M\times M}^{\alpha} - \mathrm{Q}_{M\times M}^{(\alpha,\phi)}\right) \left[A\mathrm{G}_{M\times M}^{(\nu)} + B\mathrm{G}_{M\times M}^{(\omega)} + CI\right] \\ &+\mathrm{F}_M^1 \left(A\mathrm{G}_{M\times M}^{(\nu)} + B\mathrm{G}_{M\times M}^{(\omega)} + CI\right) - \mathrm{F}_M^2 = 0. \end{aligned} \tag{41}$$

Which is a simple algebraic equation. Solving this equation for $\mathbf{K}_M$ and putting it in (39) we get the required approximate solution of the corresponding boundary value problem.

4.1 Coupled system of boundary value problems for fractional order differential equations

In this subsection, we use operational matrices to derive procedure for the numerical solutions of coupled system. We consider the following general coupled system of FDEs as

$$\begin{aligned} D^{\alpha}x(t) + A_1 D^{\nu_1}x(t) + B_1 D^{\nu_2}y(t) + C_1 x(t) + D_1 y(t) = f(t), \ \ t \in [0,T], \\ D^{\beta}y(t) + A_2 D^{\omega_1}x(t) + B_2 D^{\omega_2}y(t) + C_2 x(t) + D_2 y(t) = g(t), \ \ t \in [0,T], \end{aligned} \quad (42)$$

subject to the boundary conditions given by

$$\begin{aligned} x(0) = x_0, \ \ x(T) = x_1, \\ y(0) = y_0, \ \ y(T) = y_1, \end{aligned} \quad (43)$$

where $1 < \alpha, \beta \leq 2, \ \ 0 < \nu_i, \omega_i < 1 (i = 1, 2)$. Further, $f(t), g(t) \in C[0, T]$ are the source term of the system (42), (43) and x_0, y_0, x_1, y_1 are real constants, $A_i, B_i, C_i, D_i (i = 1, 2)$ are any constants. To approximate the above system in terms of shifted Jacobi polynomials, we take

$$D^{\alpha}x(t) = K_M^T \Phi_M(t) \quad (44)$$

and

$$D^{\beta}y(t) = L_M^T \Phi_M(t) \quad (45)$$

Applying I^{α}, I^{β} and lemma (1) and boundary conditions, the above Eqs. (44), (45) imply that

$$x(t) = K_M^T \left(H_{M\times M}^{(\alpha)} - Q_{M\times M}^{(\alpha,\phi)} \right) \Phi_M(t) + F_M^1 \Phi_M(t) \quad (46)$$

$$y(t) = LT_M \left(H_{M\times M}^{(\beta)} - Q_{M\times M}^{(\beta,\phi)} \right) \Phi_M(t) + F_M^2 \Phi_M(t). \quad (47)$$

Now taking fractional order derivative of (46), (47), we have

$$\begin{aligned} D^{\nu_1}x(t) = K_M^T \left(H_{M\times M}^{(\alpha)} - Q_{M\times M}^{(\alpha,\phi)} \right) G_{M\times M}^{(\nu_1)} \Phi_M(t) + F_M^1 G_{M\times M}^{(\nu_1)} \Phi_M(t) \\ D^{\omega_1}x(t) = K_M^T \left(H_{M\times M}^{(\alpha)} - Q_{M\times M}^{(\alpha,\phi)} \right) G_{M\times M}^{(\omega_1)} \Phi_M(t) + F_M^1 G_{M\times M}^{(\omega_1)} \Phi_M(t) \end{aligned} \quad (48)$$

and

$$\begin{aligned} D^{\nu_2}y(t) = L_M^T \left(H_{M\times M}^{(\beta)} - Q_{M\times M}^{(\beta,\phi)} \right) G_{M\times M}^{(\nu_2)} \Phi_M(t) + F_M^2 G_{M\times M}^{(\nu_2)} \Phi_M(t) \\ D^{(\omega_2)}y(t) = L_M^T \left(H_{M\times M}^{(\beta)} - Q_{M\times M}^{(\beta,\phi)} \right) G_{M\times M}^{(\omega_2)} \Phi_M(t) + F_M^2 G_{M\times M}^{(\omega_2)} \Phi_M(t). \end{aligned} \quad (49)$$

Putting (45)–(49) together with the use of approximation $f(t) \simeq F_M^3 \Phi_M(t), \ \ g(t) \simeq F_M^4 \Phi_M(t)$ in (42), we obtain

$$\begin{aligned} K_M^T \Phi_M(t) + A_1 \left[K_M^T \left(H_{M\times M}^{(\alpha)} - Q_{M\times M}^{(\alpha,\phi)} \right) G_{M\times M}^{(\nu_1)} \Phi_M(t) + F_M^1 G_{M\times M}^{(\nu_1)} \Phi_M(t) \right] \\ + B_1 \left[L_M^T \left(H_{M\times M}^{(\beta)} - Q_{M\times M}^{(\beta,\phi)} \right) G_{M\times M}^{(\nu_2)} \Phi_M(t) + F_M^2 G_{M\times M}^{(\nu_2)} \Phi_M(t) \right] \\ + C_1 \left[K_M^T \left(H_{M\times M}^{(\alpha)} - Q_{M\times M}^{(\alpha,\phi)} \right) \Phi_M(t) + F_M^1 \Phi_M(t) \right] \\ + D_1 \left[L_M^T \left(H_{M\times M}^{(\beta)} - Q_{M\times M}^{(\beta,\phi)} \right) \Phi_M(t) + F_M^2 \Phi_M(t) \right] \\ - F_M^3 \Phi_M(t) = 0 \end{aligned} \quad (50)$$

$$\begin{aligned}
&\mathrm{L}_M^T\Phi_M(t) + A_2\left[\mathrm{K}_M^T\left(\mathrm{H}_{M\times M}^{(\alpha)} - \mathrm{Q}_{M\times M}^{(\alpha,\phi)}\right)\mathrm{G}_{M\times M}^{(\omega_1)}\Phi_M(t) + \mathrm{F}_M^1\mathrm{G}_{M\times M}^{(\omega_1)}\Phi_M(t)\right]\\
&\quad + B_2\left[\mathrm{L}_M^T\left(\mathrm{H}_{M\times M}^{(\beta)} - \mathrm{Q}_{M\times M}^{(\beta,\phi)}\right)\mathrm{G}_{M\times M}^{(\omega_2)}\Phi_M(t) + \mathrm{F}_M^2\mathrm{G}_{M\times M}^{(\omega_2)}\Phi_M(t)\right]\\
&\quad + C_2\left[\mathrm{K}_M^T\left(\mathrm{H}_{M\times M}^{(\alpha)} - \mathrm{Q}_{M\times M}^{(\alpha,\phi)}\right)\Phi_M(t) + \mathrm{F}_M^1\Phi_M(t)\right]\\
&\quad + D_2\left[\mathrm{L}_M^T\left(\mathrm{H}_{M\times M}^{(\beta)} - \mathrm{Q}_{M\times M}^{(\beta,\phi)}\right)\Phi_M(t) + \mathrm{F}_M^2\Phi_M(t)\right]\\
&\quad - \mathrm{F}_M^4\Phi_M(t) = 0.
\end{aligned}$$

Rearranging the terms in above system (50) and using the notation for simplicity

$$\begin{aligned}
&\mathrm{S} = \mathrm{F}_M^1\left(A_1\mathrm{G}^{(\nu_1)} + C_1 I\right) + \mathrm{F}_M^2\left(B_1\mathrm{G}^{(\nu_2)} + D_1 I\right) - \mathrm{F}_M^3\\
&\mathrm{R} = \mathrm{F}_M^1\left(A_2\mathrm{G}^{(\omega_1)} + C_2 I\right) + \mathrm{F}_M^2\left(B_2\mathrm{G}^{(\omega_2)} + D_2 I\right) - \mathrm{F}_M^4\\
&\mathrm{H}^{(\alpha)} - \mathrm{Q}^{(\alpha,\phi)} = \widehat{P}, \quad \mathrm{H}^{(\beta)} - \mathrm{Q}^{(\beta,\phi)} = \widehat{Q}
\end{aligned} \tag{50a}$$

then witting in matrix form, we have

$$\begin{aligned}
&\left[\mathrm{K}_M^T\ \mathrm{L}_M^T\right]\begin{bmatrix}\Phi_M(t) & 0\\ 0 & \Phi_M(t)\end{bmatrix}\\
&+\left[\mathrm{K}_M^T\ \ \mathrm{L}_M^T\right]\begin{bmatrix}\widehat{\mathrm{P}}\left(A_1\mathrm{G}^{\nu_1} + C_1 I\right) & \widehat{\mathrm{Q}}\left(B_1\mathrm{G}^{\nu_2} + D_1 I\right)\\ \widehat{\mathrm{P}}\left(A_2\mathrm{G}^{\omega_1} + C_2 I\right) & \widehat{\mathrm{Q}}\left(B_2\mathrm{G}^{\omega_2} + C_2 I\right)\end{bmatrix}\begin{bmatrix}\Phi_M(t) & 0\\ 0 & \Phi_M(t)\end{bmatrix}\\
&+\left[\mathrm{S}\ \ \mathrm{R}\right]\begin{bmatrix}\Phi_M(t) & 0\\ 0 & \Phi_M(t)\end{bmatrix} = 0\\
&\Rightarrow \left[\mathrm{K}_M^T\ \ \mathrm{L}_M^T\right] + \left[\mathrm{K}_M^T\ \ \mathrm{L}_M^T\right]\begin{bmatrix}\widehat{\mathrm{P}}\left(A_1\mathrm{G}^{\nu_1} + C_1 I\right) & \widehat{\mathrm{Q}}\left(B_1\mathrm{G}^{\nu_2} + D_1 I\right)\\ \widehat{\mathrm{P}}\left(A_2\mathrm{G}^{\omega_1} + C_2 I\right) & \widehat{\mathrm{Q}}\left(B_2\mathrm{G}^{\omega_2} + C_2 I\right)\end{bmatrix} + \left[\mathrm{S}\ \ \mathrm{R}\right] = 0
\end{aligned} \tag{51}$$

which is an algebraic equation and can easily be solved for unknown matrices $\mathrm{K}_M, \mathrm{L}_M$, then putting it in (38) and (39), we get required approximations for $x(t), y(t)$.

5. Numerical examples

In this section, we will apply our proposed scheme to some practical problems.

Example 1. *Consider the problems of Buckling of a thin vertical column such that when a constant compressive force F is applied to the said thin column of length L hanged at both ends whose general differential equations in the absence of source term are given by*

$$\begin{aligned}
&EL\frac{d^\alpha y}{dx^\alpha} + Fy = f(t), \quad 1 < \alpha \le 2,\\
&\textit{subject to the boundary conditions} \quad y(0) = y_1, \ y(L) = y_2,
\end{aligned} \tag{52}$$

M	a	b	$\lvert y_{app} - y_{exact}\rvert$	a	b	$\lvert y_{app} - y_{exact}\rvert$
2	0	0	8×10^{-2}	1	1	4×10^{-2}
4	0	0	6×10^{-3}	1	1	3×10^{-3}
6	2	2	1.2×10^{-3}	1	1	2×10^{-3}
8	0.5	0.5	8×10^{-4}	1.5	1.5	2×10^{-4}
8	0	0	1×10^{-5}	−0.5	−0.5	2×10^{-5}
10	2	0	2.3×10^{-5}	0	2	4×10^{-5}

Table 1.
Maximum absolute error for $T = 1, \alpha = 1.8, \nu_1 = \nu_2 = 0$ for example, (1).

where E, I are Young modulus and moment of inertia, respectively. Where $E = 2.6 \times 10^7 lb/in, I = 0.25\pi^2 R^4$, where R is the radius of the column. Taking $L = 1in, R = 1in$ the exact deflection at $\alpha = 2$ is $y(x) = \sin(\pi x)$. We apply our technique introduced in section (4) for various choices of a, b, M. The observations are given in **Table 1** for the absolute error $\lvert y_{app} - y_{exact}\rvert$.

In **Table 2**, we have shown the maximum absolute error for various choices of a, b, M. Also in **Figures 1** and **2**, we have shown the images of their comparison between exact and approximate values. Further, we observed that as the $\alpha \to 2$, the error is reducing and one can check that at $\alpha = 2$, the absolute error is below 1.8×10^{-13} as given in the next table. From above tables, we observe that as order $\alpha \to 2$

M	α	ν_1	ν_2	$\lvert Y_{app} - Y_{exact}\rvert$
4	2	1	1	1×10^{-3}
6	2	1	1	7×10^{-17}
	1.8	0.9	0.8	2×10^{-4}
8	1.7	1	0.8	2×10^{-4}
10	1.9	0.999	0.999	3×10^{-5}

Table 2.
Maximum absolute error at $a = b = 0$, and for different α, ν_1, ν_2 for example, (2).

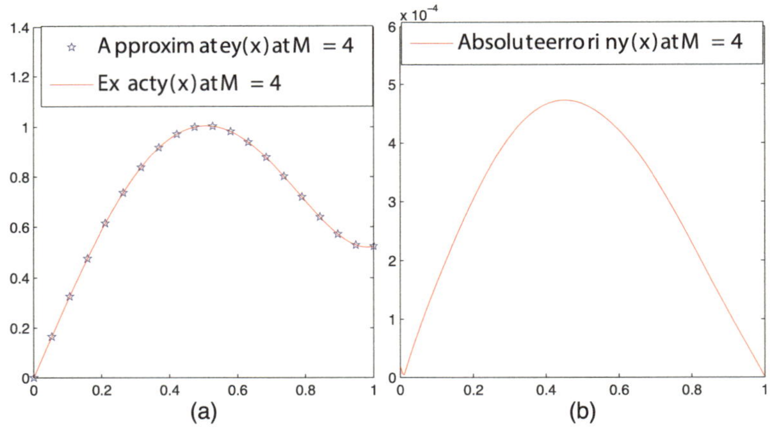

Figure 1.
Subplot (a) represents comparison of approximate and exact $y(x)$ of example 1. Stars and dots represent exact solution while the red curve represents approximate solution. Setting $\alpha = 1.8, M = 4, T = 1, a = b = 1$. the subplot represents their absolute error at $M = 8$.

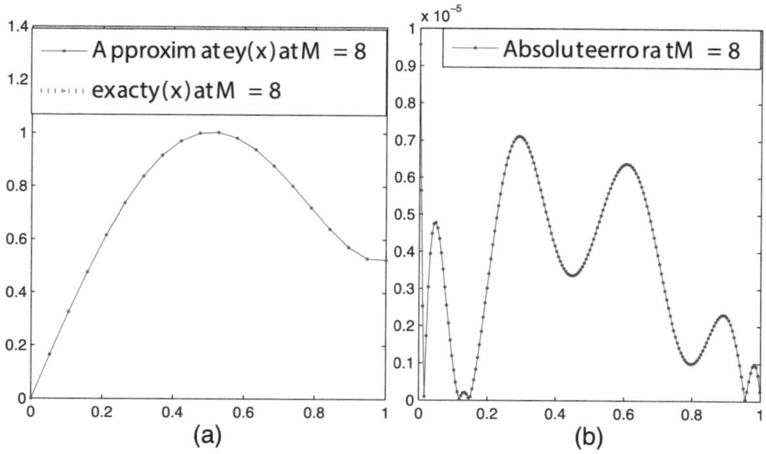

Figure 2.
Subplot (a) represents comparison of approximate and exact y(x) of example 1.Red dots represent exact solution while the blue curve represents approximate solution. Setting $\alpha = 1.8, a = b = 0, M = 8, T = 1$. the subplot (b) represents their absolute error at $M = 8$.

the approximation converges to the value at $\alpha = 2$. This phenomenon shows the accuracy of the approximate solutions.

Example 2. *Consider the following general problem*

$$\begin{aligned} &D^{\alpha}Y + AD^{\nu_1}Y + BD^{\nu_2}Y + CY = f(x), \quad 1 < \alpha \leq 2, \quad 0 < \nu_1, \nu_2 \leq 1, \\ &\textit{subject to the boundary conditions} \qquad Y(0) = 1, \quad Y(1) = 5. \end{aligned} \tag{53}$$

Let the exact solution $Y(x) = x^4 + x^3 + x^2 + x + 1$, and $A = 1, B = 2, C = 3$, the source terms

$$\begin{aligned} f(x) = {} & 3x^4 + 3x^3 + 15x^2 + 9x + 5 + \frac{24}{\Gamma\frac{21}{5}}x^{\frac{16}{5}} + \frac{6}{\Gamma\frac{16}{5}}x^{\frac{11}{5}} + \frac{2}{\Gamma\frac{11}{5}}x^{\frac{6}{5}} + \frac{1}{\Gamma\frac{6}{5}}x^{\frac{1}{5}} \\ & + \frac{48}{\Gamma\frac{23}{5}}x^{\frac{18}{5}} + \frac{12}{\Gamma\frac{18}{5}}x^{\frac{13}{5}} + \frac{4}{\Gamma\frac{13}{5}}x^{\frac{8}{5}} + \frac{2}{\Gamma\frac{8}{5}}x^{\frac{3}{5}}. \end{aligned} \tag{54}$$

We approximate the solutions of this problem with proposed method by setting $\alpha = 2, \nu_1 = 0.8, \nu_2 = 0.4, a = b = 1, M = 8, T = 1$. and observed that the scheme provides much more accurate approximations for very small-scale level even at scale level $M = 8$ maximum absolute error is 9×10^{-17} as shown in **Figure 3**.

The absolute error for different choices of a, b and M and for $\alpha = 2, \nu_1 = 0.8, \nu_2 = 0.4$ is given in **Table 3**.

From **Table 4**, we observed that as the orders of derivatives tend to their corresponding integer values, the approximate solutions tend to its exact value, which demonstrate the accuracy of numerical solutions obtained in our proposed method.

In the following examples, we solve some coupled systems under the given conditions by our proposed method.

Example 3. Consider the coupled system given by

$$\begin{aligned} &D^{\alpha}X + A_1D^{\nu_1}X + B_1D^{\nu_2}Y + C_1X + D_1Y = f(x), \quad 1 < \alpha \leq 2, \quad 0 < \nu_1, \nu_2 \leq 1, \\ &D^{\beta}Y + A_2D^{\omega_1}X + B_2D^{\omega_2}Y + C_2X + D_2Y = g(x), \quad 1 < \beta \leq 2, \quad 0 < \omega_1, \omega_2 \leq 1, \end{aligned} \tag{55}$$

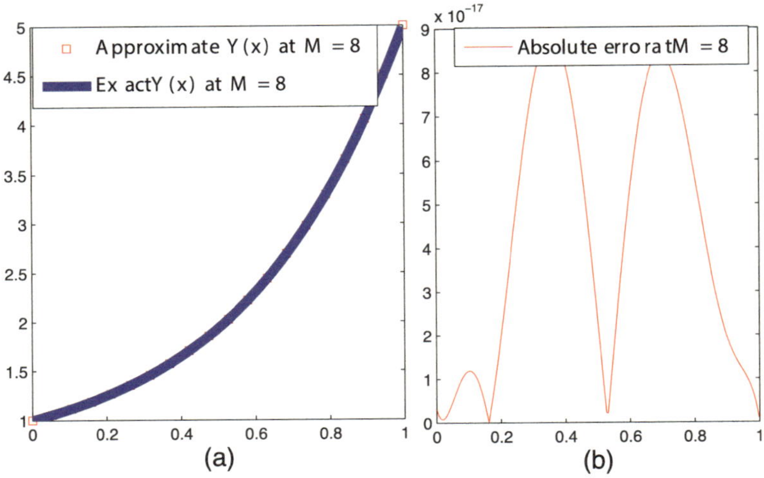

Figure 3.
Subplot (a) represents comparison of approximate and exact $Y(x)$ of example 2. Blue curve represents exact solution while the red square dots represent approximate solution. The subplot (b) represents their absolute error at $M = 8$.

M	a	b	α	$\lvert y_{app} - y_{exact} \rvert$
4	0	0	1.7	4×10^{-3}
6			1.8	4×10^{-5}
	1	1	1.9	1.5×10^{-4}
			2	3.5×10^{-13}
6	0	0	2	1.8×10^{-13}

Table 3.
Maximum absolute error at various values of α, a, b and M for example, (1).

M	a	b	$\lvert Y_{app} - Y_{exact} \rvert$
4	0	0	1×10^{-3}
6	0.5	0.5	1.6×10^{-15}
8	1	1	9×10^{-17}
8	0.5	−0.5	2.5×10^{-15}
10	0	2	4×10^{-16}

Table 4.
Maximum absolute error at $\alpha = 2, \nu_1 = 0.8,\ \nu_2 = 0.4$ for example, (2).

subject to the boundary conditions

$$\begin{aligned} X(0) = 0.5, \quad X(1) = 0.5 \\ Y(0) = 0.6, \quad Y(1) = 0.6. \end{aligned} \tag{56}$$

We solve this problem under the given parameters

$$\begin{aligned} &A_1 = 2, A_2 = 2, B_1 = 2, B_2 = 2, C_1 = 3, C_2 = 4, D_1 = 5, D_2 = 5, \\ &\alpha = \beta = 2, \nu_1 = \omega_2 = 1, \nu_1 = \omega_1 = 1, a = b = 0. \end{aligned}$$

Let the exact solutions at $\alpha = \beta = 2, \nu_i = \omega_i = 1 (i = 1, 2)$ are given by

$$X(x) = x^2(T-x)^2 + 0.5, Y(x) = x(T-x)^3 + 0.6, \tag{57}$$

the source terms are given by

$$f(x) = 4x(2x-2) - 6x(x-1)^2 - 5x(x-1)^3 + 2(x-1)^2 - 2(x-1)^3 + 3x^2(x-1)^2 + \frac{\left(3172393274221175x^{\left(\frac{7}{5}\right)}(50x^2 - 85x + 34)\right)}{33495522228568064} + 2x^2 + \frac{9}{2} \tag{58}$$

$$g(x) = 2x^2(2x-2) - 2x(x-1)^2 - 5x(x-1)^3 - 6(x-1)^2 - 2(x-1)^3 - 3x(2x-2) + 3x^2(x-1)^2 + \frac{9}{2} \tag{59}$$

We approximate the coupled systems at scale level $M = 6$ for the given parameters. We see from **Figures 4** and **5** that our method works very well, and the

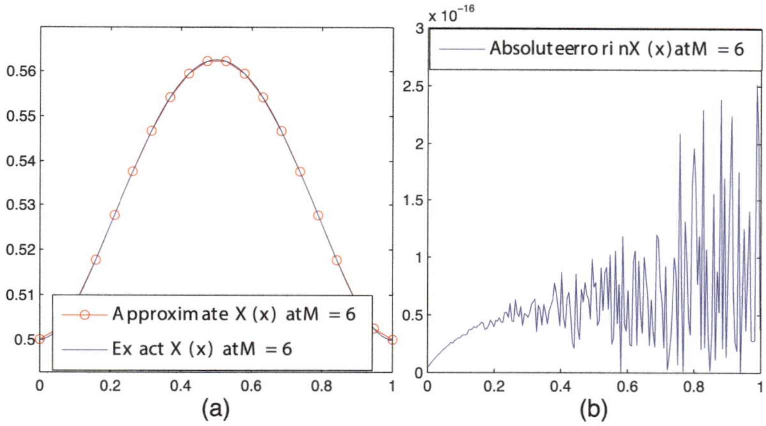

Figure 4.
Subplot (a) represents comparison of approximate and exact $X(x)$ of example 3. Blue curve represents exact solution while the red circles represent approximate solution. The subplot (b) represents their absolute error at $M = 6$.

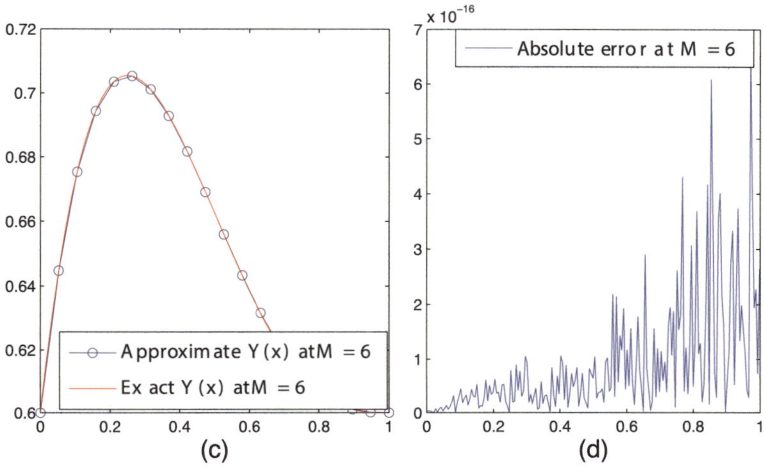

Figure 5.
Subplot (c) represents comparison of approximate and exact $Y(x)$ of example 3. Blue dots represent exact solution while the red curve represents approximate solution. The subplot (d) represents their absolute error at $M = 6$.

M	a	b	$\|X_{app} - X_{exact}\|$	$\|Y_{app} - Y_{exact}\|$
4	0	0	1×10^{-3}	1×10^{-3}
6			3×10^{-16}	7×10^{-16}
6	1	1	4.5×10^{-16}	7×10^{-16}
8	0.5	1	4.5×10^{-16}	6×10^{-16}
8	1	0.5	3.5×10^{-16}	1.2×10^{-15}

Table 5.
Maximum absolute error at various values of a, b and M for example, (3).

M	α	β	ν_1	ν_2	ω_1	ω_2	$\|X_{app} - X_{exact}\|$	$\|Y_{app} - Y_{exact}\|$
4	1.5	1.5	0.5	0.5	0.5	0.5	2.5×10^{-3}	1×10^{-3}
6	1.6	1.6	0.6	0.6	0.6	0.6	3×10^{-3}	2×10^{-3}
6	1.7	1.7	0.7	0.7	0.7	0.7	6×10^{-4}	1.2×10^{-3}
8	1.8	1.8	0.8	0.8	0.8	0.8	4.5×10^{-4}	1×10^{-4}
8	1.9	1.9	0.9	0.9	0.9	0.9	9×10^{-5}	2×10^{-5}
10	2	2	1	1	1	1	4.5×10^{-16}	6×10^{-16}

Table 6.
Maximum absolute error at specific values of a = b = 0 for example, (3).

absolute error is about 3×10^{-16} for $X(x)$ and for $Y(x)$ this amount is 7×10^{-16}. Which is very small numbers, which prove the applicability of our methods. Absolute error for various a, b, M is given in next **Table 5**.

Absolute error for specific a, b and for various $\alpha, \beta, \nu_1, \nu_2, \omega_1, \omega_2$ is given in **Table 6**.

From **Table 5**, we see the effect of a, b at various scale level M. While **Table 6** indicates the effect of various choices of $\alpha, \beta, \nu_1, \nu_2, \omega_1, \omega_2$ at different scale level. As the scale level increases, values of $\alpha, \beta, \nu_1, \nu_2, \omega_1, \omega_2$ tend to their integer values. The absolute error is reducing and the solutions are tending to their exact value, which demonstrate the applicability of the proposed method.

Example 4. Consider the mathematical models of fractionally damped coupled system of spring masses system whose model is given by

$$\begin{aligned} &D^{\alpha}X + \frac{\lambda}{m_1}D^{\nu_1}X - \frac{\mu}{m_1}D^{\nu_2}Y + \frac{\kappa_1}{m_1}X - \frac{\kappa_2}{m_1}Y = f(x), \ \ 1 < \alpha \le 2, \ \ 0 < \nu_1, \nu_2 \le 1, \\ &D^{\beta}Y - \frac{\lambda}{m_2}D^{\omega_1}X + \frac{\lambda+\mu}{m_2}D^{\omega_2}Y - \frac{\kappa_1}{m_2}X + \frac{\kappa_1+\kappa_2}{m_2}Y = g(x), \ \ 1 < \beta \le 2, \ \ 0 < \omega_1, \omega_2 \le 1, \end{aligned} \tag{60}$$

subject to the boundary conditions

$$\begin{aligned} X(0) = 0, \quad X(1) = 0 \\ Y(0) = 1, \quad Y(1) = -1. \end{aligned} \tag{61}$$

Where the source terms are given by

$$\begin{aligned} f(x) &= \cos(\pi x) - \sin(\pi x) - \pi\cos(\pi x) - \pi\sin(\pi x) - \pi^2\sin(\pi x) \\ g(x) &= \sin(\pi x) - 2\cos(\pi x) + \pi\cos(\pi x) + 2\pi\sin(\pi x) - \pi^2\cos(\pi x). \end{aligned} \tag{62}$$

Where λ, μ are damping parameters and κ_1, κ_2 are spring constants and m_1, m_2 are masses of objects and springs are hanged from both ends. We solve this problem under the following values:

$$\lambda = 2, \mu = 2, \kappa_1 = 2, \kappa_2 = 2, m_1 = m_2 = 2,$$
then calculating the coefficient of the problems as
$$A_1 = 1, A_2 = -1, B_1 = -1, B_2 = 2, C_1 = 1, C_2 = -1, D_1 = -1, D_2 = 1.$$

Let the exact solutions at $\alpha = \beta = 2, \nu_i = \omega_i = 1 (i = 1, 2)$ are given by

$$X(x) = \sin(\pi x), \qquad Y(x) = \cos(\pi x) \tag{63}$$

The above model is approximated for the solutions by our proposed methods (**Figures 6** and 7). We observed that our method provides best approximate solutions to the problems for small-scale level $M = 5$. We also find numerical solutions for fractional values of $\alpha, \beta, \nu_i, \omega_i (i = 1, 2)$. We observe that as these orders tend to

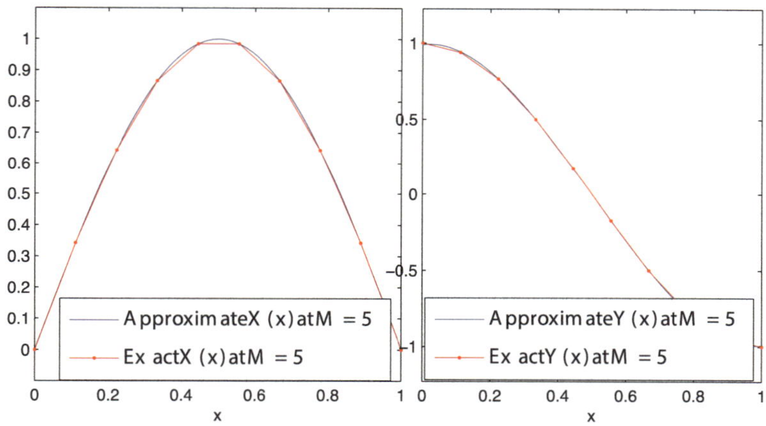

Figure 6.
Subplot (a) represents comparison of approximate and exact $X(x)$ at $M = 5$. The subplot (b) represents comparison of approximate and exact $Y(x)$ at $M = 5$ for example, (4). Blue curve represents exact solution while the red curve represents approximate solution for $a = b = 0$.

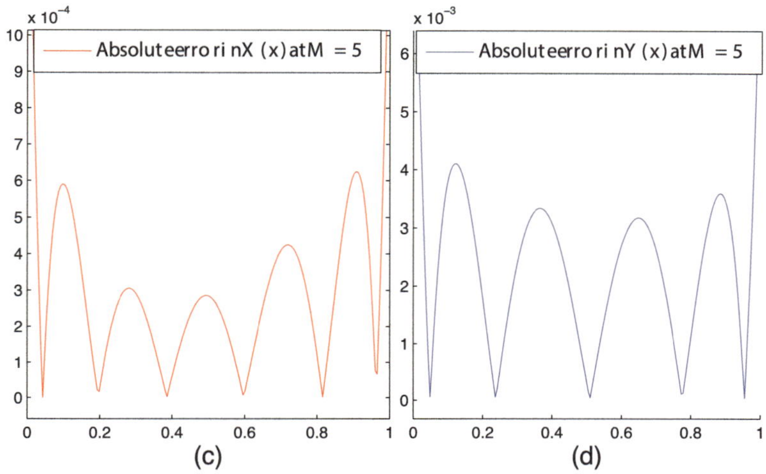

Figure 7.
Subplot (c) represents absolute error $|X_{app} - X_{exact}|$, and subplot (d) represents their absolute error $|Y_{app} - Y_{exact}|$ at $M = 6$ for example, (4) for $a = b = 0$.

M	a	b	$\lvert X_{app} - X_{exact} \rvert$	$\lvert Y_{app} - Y_{exact} \rvert$
4	0	0	1×10^{-3}	1×10^{-3}
5			10×10^{-4}	6×10^{-3}
6			9×10^{-7}	9×10^{-6}
6	1	1	4×10^{-3}	7×10^{-3}
8	0.5	.5	2×10^{-3}	1×10^{-3}
8	1	0.5	3.5×10^{-3}	1.2×10^{-3}

Table 7.
Maximum absolute error at various values of a, b and M for example, (4).

M	α	β	ν_1	ν_2	ω_1	ω_2	$\lvert X_{app} - X_{exact} \rvert$	$\lvert Y_{app} - Y_{exact} \rvert$
4	1.5	1.5	0.5	0.5	0.5	0.5	2.5×10^{-2}	1×10^{-2}
6	1.6	1.6	0.6	0.6	0.6	0.6	4×10^{-3}	2×10^{-3}
6	1.7	1.7	0.7	0.7	0.7	0.7	1×10^{-3}	1×10^{-3}
8	1.8	1.8	0.8	0.8	0.8	0.8	4.5×10^{-4}	1×10^{-4}
8	1.9	1.9	0.9	0.9	0.9	0.9	1.2×10^{-5}	2×10^{-5}
10	2	2	1	1	1	1	9×10^{-7}	8×10^{-6}

Table 8.
Maximum absolute error at $a = b = 0$ and for various $\alpha, \beta, \nu_1, \nu_2, \omega_1, \omega_2$ for example, (4).

their integer values, the solutions are tending to the exact, which prove the applicability of our method in the following tables.

Absolute error for various a, b, M and $\alpha = \beta = 2, \nu_i = \omega_i = 1 (i = 1, 2)$ is given in **Table 7**.

We observe from **Table 7** that values of a, b play important role in the approximate solution. By giving integral value to a, b, methods give best approximate solution for the given problem. Similarly for $a = b = 0$ and for various choices of $\alpha, \beta, \nu_1, \nu_2, \omega_1, \omega_2$, the absolute errors are given in **Table 8**.

From **Table 8**, it is obvious that when the orders of the derivatives approach to their integral values, the error is reducing and the approximate solutions are converging to the exact solutions. This phenomenon indicates that the proposed method is highly accurate.

6. Conclusion and discussion

In this article, we have studied shifted Jacobi polynomials. Based on these polynomials, we recalled some already existing matrices of fractional order derivative and integrations from the literature as well as we constructed a new operational matrix corresponding to boundary conditions. By means of these operational matrices, we converted the system of fractional differential equations to simple and easily soluble system of algebraic equations. There is no need of Tau-collocation method. The simple algebraic equations were easily solved, and the result were plotted. From the plot, we observe that our proposed method is highly accurate and can be applied to a variety of problems of fractional order ordinary as well as partial differential equations. We also compared our results to the exact solutions and observed that our method gave satisfactory results. The proposed method can easily and accurately can be applied to a variety of problems of applied mathematics, engineering, and physics, etc.

Acknowledgements

The authors Aziz Khan, Kamal Shah and Thabet Abdeljawad would like to thank Prince Sultan university for the support through TAS research lab.

Mathematics Subject Classification:

26A33; 34A08; 34K37.

Author details

Kamal Shah[1,2*], Eiman[2], Hammad Khalil[3], Rahmat Ali Khan[2] and Thabet Abdeljawad[1,4]

1 Department of Mathematics, University of Malakand, Khyber Pakhtunkhwa, Pakistan

2 Department of Mathematics and Sciences, Prince Sultan University, Riyadh, Saudi Arabia

3 Department of Mathematics, University of Education Lahore, Attock, Punjab, Pakistan

4 Department of Medical Research, China Medical University, Taichung, Taiwan

*Address all correspondence to: kamalshah408@gmail.com

References

[1] Podlubny I. Fractional Differential Equations, Mathematics in Science and Engineering. New York: Academic Press; 1999

[2] Lakshmikantham V, Leela S, Vasundhara J. Theory of Fractional Dynamic Systems. Cambridge, UK: Cambridge Academic Publishers; 2009

[3] Hilfer R. Applications of Fractional Calculus in Physics. Singapore: World Scientific; 2000

[4] Rossikhin YA, Shitikova MV. Applications of fractional calculus to dynamic problems of linear and nonlinear hereditary mechanics of solids. Applied Mechanics Reviews. 1997;**50**:15-67

[5] Din A, Li Y, Khan FM, et al. On Analysis of fractional order mathematical model of Hepatitis B using Atangana-Baleanu Caputo (ABC) derivative. Fractals. 2021;**30**:2240017

[6] Lakshmikantham V, Leela S. Naguma-type uniqueness result for fractional differential equations. Non-linear Anal. 2009;**71**:2886-2889

[7] Miller KS, Ross B. An Introduction to the Fractional Calculus and Fractional Differential Equations. New York: Wiley; 1993

[8] Goodrich C. Existence of a positive solution to a class of fractional differential equations. Computers & Mathematcs with Applications. 2010;**59**: 3889-3999

[9] Din A, Li Y. The extinction and persistence of a stochastic model of drinking alcohol. Results in Physics. 2021;**28**:104649

[10] Ahmad B, Nieto JJ. Existence result for a coupled system of non linear fractional differential equations with three point boundary conditions. Computers & Mathematcs with Applications. 2009;**58**:1838-1843

[11] Din A, Li Y. Lévy noise impact on a stochastic hepatitis B epidemic model under real statistical data and its fractal-fractional Atangana-Baleanu order model. Physica Scripta. 2021;**96**(12): 124008

[12] Ray SS, Bera RK. Solution of an extraordinary differential equation by Adomian decomposition method. Journal of Applied Mathematics. 2004; **4**:331-338

[13] Hashim I, Abdulaziz O, Momani S. Homotopy analysis method for fractional IVPs. Commun. Nonl. Sci. Numer. Simul. 2009;**14**:674-684

[14] Din A, Li Y. Stationary distribution extinction and optimal control for the stochastic hepatitis B epidemic model with partial immunity. Physica Scripta. 2021;**96**(7):074005

[15] Khalil H, Khan RA. A new method based on Legendre polynomials for solutions of the fractional two-dimensional heat conduction equation. Computers & Mathematics with Applications. 2014;**67**:1938-1953. DOI: 10.1016/j.camwa.2014.03.008

[16] Doha EH, Bhrawy AH, Ezz-Eldien SS. Efficient Chebyshev spectral methods for solving multi-term fractional orders differential equations. Applied Mathematical Modelling. 2011; **35**(12):5662-5672

[17] Doha EH, Bhrawy AH, Ezz-Eldien SS. A new Jacobi operational matrix: An application for solving fractional differential equations. Applied Mathematical Modelling. 2012;**36**: 4931-4943

[18] Saadatmandi A, Dehghan M. A new operational matrix for solving fractional-order differential equations. Computers & Mathematcs with Applications. 2010;**59**(3):1326-1336

[19] Singh AK, Singh VK, Singh OP. The Bernstein operational matrix of integration. Appl. Math. Sciences. 2009; **3**(49):2427-2436

[20] Rehman M u, Khan RA. the legender wavelet method for solving fractional differential equation. Commun. Nonlin. Sci. Numer. Simul. 2011;**10**:4163-4173. DOI: 10.1016/j.cnsns.2011.01.014

[21] Doha EH, Bhrawy AH, Saker MA. On the Derivatives of Bernstein Polynomials, An application for the solution of higher even-order differential equations. Boun. Val. Prob. 2011;**16**:829543. DOI: 10.1186/1687–2770-2011-16

[22] Khalil H, Khan RA. A new method based on legender polynomials for solution of system of fractional order partial differential equation. International Journal of Computer Mathematics. 2014;**67**:1938-1953. DOI: 10.1080/00207160.2014.880781

[23] Tan B, Boyd JP. Stability and long time evolution of the periodic solutions of the two coupled nonlinear Schrödinger equations. Chaos. Soli. Frac. 2001;**12**:721-734

[24] Radhakrishnan R, Sahadevan R, Lakshmanan M. Integrability and singularity structure of coupled nonlinear Schrödinger equations. Chao. Soli. Fract. 1995;**5**:2315-2327

[25] Bakkyaraj T, Sahadevan R. On solutions of two coupled fractional time derivative Hirota equations. Nonlinear Dynamics. 2014;**77**:1309-1322

[26] Rienstra SW. Non-Linear free vibrations of coupled spans of overhead transmission lines. In: Manley J et al., editors. Proceedings of the Third European Conference on Mathematics in Industry, August 27–31 1990 Glasgow. Alphen aan den Rijn, Netherlands: Kluwer Academic Publishers and B.G. Teubner Stuttgart; 1990. pp. 133-144

[27] Din A et al. Caputo type fractional operator applied to Hepatitis B system. Fractals. 2021;**31**:2240023

[28] Bhrawy AH, Alofi AS. A Jacobi-Gauss collocation method for solving nonlinear Lane-Emden type equations. Commun Nonlinear. Sci. Numer. Simulat. 2012;**17**:62-70. DOI: 10.1016/j.cnsns.2011.04.025

[29] Kilbas AA, Srivastava H, Trujillo J. Theory and application of fractional differential equations, North Holland Mathematics Studies. Vol. 204. Amsterdam: Elseveir; 2006

[30] Doha EH, Bhrawy AH, Hafez RM. On shifted Jacobi spectral method for high-order multi-point boundary value problems. Commun. Nonl. Sci. Num. Simul. 2012;**17**(10):3802-3810

[31] Bhrawy AH, Alghamdi MA. Numerical solutions of odd order linear and nonlinear initial value problems using shifted Jacobi spectral approximations. Abst. Appl. Anal. 2012; **2012**:364-360